The Four Realms of Existence

The Four Realms of Existence

A NEW THEORY OF BEING HUMAN

Joseph E. LeDoux

THE BELKNAP PRESS OF
HARVARD UNIVERSITY PRESS
Cambridge, Massachusetts
London, England
2023

Printed in the United States of America

First printing

Library of Congress Cataloging-in-Publication Data
Names: LeDoux, Joseph E., author.
Title: The four realms of existence : a new theory of being human / Joseph E. LeDoux.
Description: Cambridge, Massachusetts ; London, England : The Belknap Press of Harvard University Press, 2023. | Includes bibliographical references and index.
Identifiers: LCCN 2023000719 | ISBN 9780674261259 (cloth)
Subjects: LCSH: Cognitive neuroscience. | Human beings. | Self. | Biology. | Neurobiology. | Cognition. | Consciousness.
Classification: LCC QP360.5 .L43 2023 | DDC 612.8/233—dc23/eng/20230302
LC record available at https://lccn.loc.gov/2023000719

To Nancy, and our first forty-three years together

I made a map of your mind, I've charted my course

I'm sailing deep inside, I've got the winds of force

Got the heat of your heart, to keep me from the cold

I got the currents of will, to take me to your soul

—"Map of Your Mind," The Amygdaloids

Contents

Preface

I conceived this book in August 2020 while I was sheltered in the Catskills in upstate New York, riding out the peak of the COVID-19 pandemic. I moved the manuscript into the home stretch back in Brooklyn in the summer of 2022, tidied up while I was a visiting scholar at the London School of Economics that fall, and submitted it Thanksgiving weekend. Somehow it seems fitting that, after two-and-a-half squeaky-clean years, I finally came down with COVID just as this book went into production.

I don't have a writing schedule. I love writing and just do it as often as I can. This is how I find out what is on my mind. Regardless of whether you are a methodical writer who thoroughly researches your topic, or a leap-before-you-look writer who, like me, has an idea and figures things out along the way, the content that results is not generated by you explicitly planning each word, phrase, or conception.

As this book unfolds, you will see that this is a key part of the story I will tell—that your conscious understanding of your mind is an interpretation or narration that flows effortlessly from non-conscious processes. I'm not referring here to the deep, dark, Freudian unconscious. I have something far more mundane in mind, something I borrowed from the pioneering psychologist and brain researcher Karl Lashley, who said that every conscious thought is preceded by—that is, is based on—non-conscious, or better yet, pre-conscious, cognitive processing.

While this book is partly about how consciousness depends on non-conscious processing in the brain, I tell a much bigger story, one in which our biological makeup is the foundation of our neural makeup, which underlies our cognitive makeup and enables our conscious makeup. No, it's not about reducing consciousness to cognition, cognition to the brain, and the brain to biology. It is instead about how everything that can be said about you as a living thing is subsumed by interactions between biological, neurobiological, cognitive, and conscious ways of being. These are our realms of existence.

All living things, all organisms, exist biologically. But some of these, namely animals, evolved nervous systems, and so they also exist neurobiologically. Of these, some can think and plan, and thus exist cognitively. Finally, some cognitive organisms also exist consciously. All that we are is subsumed within these entwined realms of existence.

Much thanks to my wife, Nancy Princenthal, a fantastic writer and editor. With her keen eye and sharp mind she made invaluable contributions. Our son, Milo LeDoux, also helped by posing challenging questions on key topics, like animal consciousness and conscious AI. It was great working with Robert Lee again. He made the illustrations for my earlier book *Anxious*, and also created the original figures in this book.

I am extremely grateful to colleagues who have spent time talking with me about or reading drafts of this book. Included are Tyler Volk, Richard Brown, Matthias Michel, Hakwan Lau, Steve Fleming, Kenneth Schaffner, Nick Shea, David Rosenthal, Owen Flanagan, Frédérique de Vignemont, Karin Roelofs, and Max Bennett. Others advised me on specific points, including Kalina Christoff, Nathaniel Daw, Bernard Balleine, Niki Clauton, Songyao

Ren, Peter Mineke, Paul Cisek, Todd Preuss, Steve Weiss, Jon Kaas, Charan Ranganath, Onur Güntürkün, Brian Key, and Deborah Brown.

It has been a pleasure getting to know, and to work with, Andrew Kinney at Harvard University Press. His insights along the way, and his pressure on me to shorten the manuscript, made the book much better. Kathleen Drummy and Christine Thorsteinsson at the Press were very helpful with the process. Copyeditor Julie Carlson did a fantastic job, but beyond that, she was a delight to work with. As always, my agent since 1996, Katinka Matson at Brockman Inc., has been by my side, start to finish.

New York University has been a wonderful academic home to me since 1989. The university, and my department, The Center for Neural Science, have supported me in innumerable ways, making it possible for me to be a scientist and a writer, as well as a musician.

William Chang has been my personal assistant for more than two decades. He has made everything possible. Claudia Farb and Mian Hou, two other multi-decade staff members, have also helped in ways that defy words. A special shout out to Mian for his work on my various websites, including the one for this book—https://joseph-ledoux.com/The-Four-Realms-of-Existence.html.

Writing a book about science means that you have to have some science to write about. But unlike my first three books (*The Emotional Brain, Synaptic Self,* and *Anxious*), my most recent two (*The Deep History of Ourselves* and this one) are more conceptual. That does not mean the work in my lab had nothing to do with them. None of my books would have been possible without the empirical findings obtained by the many fantastic researchers I have had the pleasure of sharing my lab with over the decades. Thank you, one and all.

The Four Realms of Existence

Introduction

Who Are You?

One afternoon in 1976 I was sitting in a small camper trailer in Bennington, Vermont, with a teenager named Paul. I said "who" aloud, and simultaneously pushed a button that caused the words "are you?" to briefly appear on the left side of a projection screen. The boy's left hand reached forward toward a pile of Scrabble letters. He pulled out four and arranged them to spell "Paul."

Studies of Paul, or PS as he is known in the literature, were part of my PhD research at the State University of New York at Stony Brook under the mentorship of Michael Gazzaniga. Mike, just ten years my elder, had achieved scientific stardom in the 1960s for his PhD work at Caltech on so-called split-brain patients. People received this surgery because available medications were unable to relieve their epilepsy. By severing the nerve fibers that connected the two hemispheres of their brains, most notably the bundle of nerves called the corpus callosum, the seizures became more manageable.

When I first arrived at Stony Brook in the fall of 1974, Mike was not yet my mentor. That transpired after a fellow student showed me an article that Mike had published a couple of years earlier with the title: "One Brain—Two Minds?" I was intrigued, and had the good fortune of being able to join Mike's lab. A new group of patients was being given the "split-brain" operation by a neurosurgeon at Dartmouth Medical School, and I got in at just the right time. During my first year, Mike bought the camper mentioned earlier, and we converted it into a mobile testing lab. Over the next several years, a couple of other students and I made frequent trips to New England with Mike to test these patients.

In his PhD studies, Mike had showed that when the brain was split, information presented to the right hemisphere stays there. This was achieved by instructing the patient to fixate his or her eyes on a small dot in the center of a projection screen. Because of the way the nerves from the eye to the brain are routed, stimuli on the left side of the screen are directed to the right hemisphere and stimuli to the right to the left hemisphere. And because language is preferentially represented in the left hemisphere, the patient cannot talk about information presented only to the right hemisphere. For example, if only the right hemisphere was shown a picture of an apple, and the patient was asked "What did you see?" the patient would say something like, "I didn't see anything," because the vocal left hemisphere did not process that information. But the left hand, which is preferentially connected with the right hemisphere, could reach into a bag with several objects and pull the apple out.

Mike's PhD research raised the possibility of two minds living side by side inside the head, one in each hemisphere. But the evidence at the time was not conclusive. Even a brief conversation with the loquacious left hemisphere gives you the sense that you are talking to a full-on human being—you hear about opinions, beliefs, memories, fears, and hopes. It was much less clear whether there was a sentient being in the right hemisphere, because exchanges with it mostly involved the left hand choosing which of several stimuli had been seen.

The study of Paul involving the Scrabble letters suggested to us that that the right hemisphere of a split-brain patient could indeed have a conscious sense of self. Paul's right hemisphere knew that he was Paul. This is not trivial. A person's name is a key marker of their identity, a hook on which hangs much of what is known about who they are.

But Paul was unusual. Unlike most split-brain patients, who have language only in their left hemisphere, Paul could understand spoken and written language with both hemispheres, despite being able to talk only via his left hemisphere. That's why we could ask his right hemisphere, "Who are you?" and get a response. Because the amount of information that can be processed in a quick flash is limited, I spoke part of the question out loud and presented the rest visually. As a result, both hemispheres heard the word "who," but only the right hemisphere saw "are you."

In other studies of Paul, we were able to use these kinds of simple probes to show that his right hemisphere could imagine the future. When asked what his goal in life was, the left hand, the behavioral agent of the right hemisphere, spelled out "race car driver." Strikingly, the left hemisphere had declared in conversations with us that it wanted to be a draftsman. That two different answers emerged from the same brain was mind-blowing.

In yet another study, we presented two pictures to Paul, each positioned on different sides of the screen, and instructed him to point to the picture that matched what he saw. For example, the left hemisphere saw a chicken claw and the right a snow scene. In response, the left hand pointed to a shovel and the right hand to a chicken. When asked why he chose those, Paul said: I saw a chicken claw so I pointed to a chicken, and you need a shovel to clean out the chicken shed. In other words, Paul's left hemisphere made up a story about his right hemisphere's action that matched the knowledge it (the left hemisphere) possessed, so that the behavior of the whole person, Paul, made sense.

That night we were at The Brasserie, a wonderful little bistro in Bennington, Vermont, sipping Jack Daniels at the bar, which was our standard way of consolidating the day's events. Our conversation led us to the idea that we all spin tales or narratives to make sense of ourselves and our world, a notion that would figure prominently in our 1978 book *The Integrated Mind*.

Ever since my studies of Paul, I have been fascinated with questions about how our conscious experiences come about in the brain, especially in relation to emotions and memory, and have summarized updates of my views in my books over the years. But the book you are reading and the one right before, *The Deep History of Ourselves: The Four-Billion-Year Story of How We Got Conscious Brains,* have a special connection. In a phrase, *Deep History* was about "The way we were"—how we evolved. And this book is about "The way we are"—what it is to be a living human.

Anatomy of a Feast

I serve this book in five courses. The first gives you a taste of what is to come. It raises questions about the value of ancient, entrenched notions of *self* and *person* as ways to define what and who we are. I reconceptualize the useful information obtained by empirical research on these topics and place it in a new framework, one based more on scientific rather than philosophical principles. In particular, I argue that everything about what it is to be a human being can be understood in terms of our biological, neurobiological, cognitive, and conscious realms of existence. These make us biological beings with nervous systems that can think, plan, and decide, and that can experience our inner thoughts and feelings.

The next four parts of the book delve deeply into these realms. I start by describing how all organisms exist biologically, but some of these (namely animals) have a nervous system and exist neurobiologically. Some animals additionally exist cognitively, and some of these also exist consciously. But my primary goal is not to identify the ways that different organisms in the tree of life exist (though there will a bit of that). My main focus is instead on us, on the four realms of existence that account for what it is to be a human being.

At the very end of the book, as the feast is winding down, I will revisit the studies of Paul, and the idea that we all try to make sense of what and who we are by narrating our experiences to ourselves and to others. But the path to, and through, that final taste involves first savoring the four realms of existence.

A final note about what you will feast on. I wrote this book hoping to connect with experts and lay readers alike. With the aim of being "user friendly," it consists of short, pithy chapters focused on specific points—though a few chapters broke the rule. You will also encounter no footnotes, endnotes, or citations as you read, other than a list of suggested readings at the back of the book. The actual citations and notes are on the book's website: https://joseph-ledoux.com/The-Four-Realms-of-Existence.html. Enjoy.

PART I

OUR REALMS OF EXISTENCE

1

What Is a Human Being?

Since ancient times, humans have thought of their bodies and minds as separate spheres of existence. The body is physical. It is the source of aches and pains. But the mind is mental. It perceives, remembers, thinks, believes, feels, and imagines.

Today, many of us understand that the mental aspect of who we are is embedded in the part of the body known as the brain, and therefore is also part of our physical, bodily existence. Still, even true believers of the physical nature of the mind sometimes feel as though it possesses some quality or qualities lacking in other physical systems within our body, and even within our brain. For example, we have firsthand knowledge of the mental states we refer to as perceptions, memories, thoughts, and emotions, but we lack knowledge of the processes in the brain that control digestion, respiration, or heart rhythm, and much of our behavior. What is it about the mental stuff that makes it seem so different from the rest of the physical stuff that constitutes who and what we are?

Just as your mind depends on your brain, your brain, being part of your body, depends on the life-sustaining functions of other components of your body. If your heart stops beating, or your lungs collapse, all your other organs, including your brain, will soon cease to function in a way that is compatible with life. Without bodily life there is no brain function, and without brain function, no mind.

How, then, out of all this biological physicality, do we each come to exist as a being that knows it was born in the past, exists now, and will someday die?

The standard approach to such questions about individuality is to focus on psychological notions, or constructs, like the self or personality. These have long guided philosophical musings, as well as scientific theories and research, about what it is to be a human being. But there is little agreement about what self and personality refer to, and even whether they refer to real entities, as opposed to just being shorthand labels for a variety of psychologically interesting phenomena. One important reason that clarity about all this is important is because it colors ideas about mental disorders and their treatment.

If We Don't Know What We Are Looking For, We Will Never Find It

Scientific discoveries over the last several decades in fields such as physics, artificial intelligence, and genetics have led to new ideas about how human biological systems work. These findings, in challenging cherished assumptions about human nature, have resulted in an epistemological vacuum. In no small part, this is because thinking about "who and what we are" has not advanced significantly beyond traditional ideas, some put forth in ancient times.

The mind was indeed long the province of philosophers. But then, with the birth of psychology in the late nineteenth century, the mind became a scientific topic. Many psychologists eagerly applied the experimental methods of physiology to the study of inner existence in humans and in other animals. Some concluded, however, that philosophical conceptions of an unobservable, mental sphere were not readily compatible with the methods of science, and they offered a simple solution—eliminate the unmeasurable, ghostly mind and make psychology about behavior. The result of this approach, so-called behaviorism, would dominate academic psychology, especially in the United States, for decades.

In the meantime, explorations of the mind were thriving outside of psychology, especially in the young field of psychiatry, where Sigmund Freud embraced the mental sphere in an effort to treat psychological disorders—how else could you possibly do that? But by mid-century, psychopharmaceuticals had emerged as a more rigorous, scientific approach to psychiatry, one that viewed psychological problems as disease states of the brain caused by bad chemistry. Much of this research was carried out by scientists from behaviorist backgrounds. The assumption was that medications for treating psychological problems in humans could be discovered by measuring their effects on behavioral responses in animals, typically in rats or mice. The quaint designation "mental disorders" was retained, but behavioral rather than mental symptoms were emphasized, and have been ever since.

Attempts to deliver treatments that were substantially better than those discovered in the 1950s and 1960s failed over and over. But the legions of scientists involved nevertheless grew, and they soldiered on. The problem, they assumed, was that they were close, but the drugs were not yet quite right pharmacologically, or they weren't yet making their way to the right part of the brain. Surely new technologies, such as gene-informed drug discovery, when combined with "smart drugs" and with new ways to image the circuitry of the brain, would close the gap. But they haven't. Why? I believe it is because measurable behavioral responses are, at best, correlates of the mental states for which mental disorders are named. The problem is less about technological limitations than about our lack of a rigorous, scientifically based understanding of what a human being is.

The digital revolution is teaching us a similar lesson. In *Artificial You: AI and the Future of Your Mind,* the philosopher Susan Schneider noted that as we come to rely more and more on technological advances in mind-brain enhancement and artificial intelligence, our poor understanding of self, consciousness, and mind may well lead to human suffering, or even threaten our survival as a species. The remarkable capabilities of ChatGPT and Bing almost instantly raised concerns as potential threats to humanity.

Our Realms of Existence

The various phenomena that have been discovered while studying constructs like *self* and *personality* have, without question, provided important insights into human nature. But what if it is our scientific understanding of who and what we are that is confused? Specifically, what if our constructs are inadequate as conceptual hooks on which to hang the empirical findings that have been discovered in their name? Because these centuries-old notions obscure as much as they reveal, maybe the phenomena would be better served by a new conceptual home, one grounded in contemporary scientific conceptions and empirical research.

While human nature has been written about from many points of view, I think I have a novel take on this perennial puzzle. I believe that a human being can be characterized as a composite of four fundamental, parallel, entwined realms of existence that reflect our evolutionary past and account for our present ways of being.

We exist within these four realms of existence—biological, neurobiological, cognitive, and conscious—in every moment of life (especially adult life). But the kind of existence contributed by each realm is different. All four are, deep down, biological. But the neurobiological realm transcends the biological, the cognitive transcends the neurobiological, and the conscious transcends the cognitive. Together, the four account for what and who we are, including those aspects of us that fall under the rubrics of the self and personality.

The composite organismic state that emerges from our four realms of existence can be thought of as an *ensemble of being.* This amalgam varies dynamically moment to moment over the course of one's life in accordance with the activities with which each realm is occupied with, at any given moment.

I am fully aware that this short description raises more questions than it answers. But I can't lay out the solution until I explain the problem. In the next several chapters, then, I take on what I see as fatal shortcomings of *self* and *personality* as accounts of what and who a person is, thereby setting the stage for a deep dive into our four realms of existence, and the ensemble of being that they engender in each of us.

2

"Self" Doubt

Many of you are likely ready to go to the mat defending the idea that we humans have a self that runs through our brain and body. But when we say we have a self, what is it that we think we have? And does "have" mean that we possess it? If so, are we both the thing that is possessed (the self) and its owner (the entity that possesses the thing)? If this sounds philosophical, there's a good reason.

Birth of the Self

The ancient Greeks are often said to have come up with the expression "know thyself." But according to Christopher Gill, "There was no obvious lexical equivalent in the Ancient Greek (or Latin) for 'the self.' The closest they came was 'person' and 'character.' 'Self' began to appear in Old English, by way of other Northern European languages, as a means of describing 'one's own person.' By the 14th century it began to be used as part of reflexive pronouns (myself, herself, himself, itself, etc.) where the subject and object of the verb in a sentence referred to the same person or thing (I blame myself; he, himself, is the one who deserves credit)."

Today the word self, at least in Western cultures, commonly refers to those features that make one a conscious being with a first-person point of view. This way of thinking began to take shape in the seventeenth century through the writings of René Descartes. A polymath, Descartes was a philosopher, mathematician, and physicist, and if psychology and neuroscience had existed, he undoubtedly would have excelled in those areas, too.

Descartes viewed the mind as an immaterial, immortal soul. It was both spiritual (connected to God) and mental (the locus of thought, including conscious thoughts about oneself). With his famous statement *cogito, ergo sum* (I think, therefore, I am), he tied personal existence to conscious thought, and equated this *consciousness* with the soul. In his scheme, the conscious mind has agency—that is, it controls the actions of its physical body, and as a result, will be held accountable on Judgment Day. Other animals exist only bodily and, lacking conscious agency, have no afterlife in either heaven or hell.

John Locke, who followed Descartes by a few decades, rocked the Christian world when he proposed that the core of an individual was not an immaterial soul but a conscious self that persisted over time. Key to Locke's position was his novel solution to the problem of how one knows they are the same person today as yesterday. In England, where Locke resided, identity was based on social status. But Locke said it instead resulted from the continuity of one's self over time. This was achieved, he proposed, by thinking back to the past—that is, by making one's self, via conscious memory, the same self now as it was then.

Locke also introduced the idea of a *constitution of man,* which he said encompasses our body, and all our behavioral and mental faculties. (The notion of an *ensemble of being* that I mentioned earlier bears some similarity to Locke's philosophical conception, but is based on rigorously characterized biological and psychological phenomena that Locke was not privy to.) Like Cicero before him, Locke allowed for the possibility of an individual having different *personas*—different ways of being, different selves, in various situations, depending on what he or she was conscious of at the time.

By the eighteenth century, conceptions of the self were becoming even more elaborate. The moralist Bishop Butler, for example, talked about two kinds of selves—one cool and settled (a feeling of the moment) and the other passionate and / or sensual (a feeling of self-interest). Along similar divisional lines, the philosopher Adam Smith, known as the father of modern economics, described how he examined his own conduct: "I divide myself, as it were, into two persons." These he referred to as the examiner and the judge. He also took the self into the social realm, viewing one's relationship to others as a function of one's relationship to one's self. Later, Alexander Bain noted that we look at our thoughts with a "warm eye," with a "tenderness for that activity" as if it was another person.

Butler also introduced a notion of the self as the core of a monitoring system that maintains the normalcy of human nature (an idea that has become popular today in cognitive science). Deviance, including mental illness and criminality, then came to be thought of as a failure to monitor and was thus a weakness of the self.

By the eighteenth century, the vocabulary of self had proliferated in everyday language with the use of terms like self-worth, self-esteem, self-conscious, self-love, self-praise, self-pride, self-contained, self-regard, self-made, self-interest, self-confidence, self-aware, self-monitoring, self-involved, self-care, and selfish. These reinforced the idea that there is some entity or thing called the self.

Descartes's immaterial soul continued the Christian tradition of treating the soul as the basis of individuality and personal responsibility. Armed with an emerging lexicon of the self, regular people developed a broader view of "themselves" as autonomous entities with independent lives and choices (entities with agency and responsibility).

The Scientific Self

Locke was less interested than Descartes in whether the self was material or immaterial. He was more concerned with how the self was constructed by consciousness. And his *self,* unlike Descartes's *soul,* had properties by which it could be known empirically like any other natural object. This idea planted the seed that the conscious self might be the subject of scientific inquiry.

But it was not until the late nineteenth century that a scientific endeavor emerged that could pursue such an idea. It was then, in Germany, that the methods of natural science, notably physiology, began to be applied to the philosophical problems of the mind, and especially to consciousness. The result was the field of experimental psychology.

The main approach taken by these scientific psychologists was the same one that philosophers had relied on in their analyses of consciousness—introspection. What the German psychologists added was a way to treat the results of self-introspections as scientific data with strict rules for both how to go about the introspective process and how to record the observations.

In the United States, one of the early pioneers of the new psychology was William James. The experimental study of consciousness was natural for him, because he had trained as both a philosopher and physiologist. James made many important and lasting contributions to psychology, most of which he documented in a massive two-volume textbook, *Principles of Psychology,* published in 1890. The book thoroughly covered psychology circa the late nineteenth century, but more important, offered many insights that later psychologists verified.

James wrote quite a bit about the self in his textbook. For example, he continued the emerging tradition of separating the observer and the thing observed; the *I* was the entity having an experience (the observer); the *me* was the entity that the experience is about (the thing that was being observed). James also pointed out that people tend to use these personal pronouns when talking about their conscious mind, but turn to impersonal designations, such as *it,* when referring to the unconscious, or to the body. The legal system today continues this tradition, with one held responsible for deliberate, conscious actions, but sometimes acquitted for behaviors that seem to have been produced without conscious intent—as in the temporary insanity defense.

For James there were four broad kinds of selves: material (not just bodily characteristics, but also possessions, such as one's family, clothes, home, bank account, and yacht); social (interactions with others, including adopting different selves in different social situations); spiritual (our intimate inner thoughts, including moral standards); and pure ego (the ability to recognize one own thoughts by their "warmth and intimacy"—recall Bain's warm eye and tender feelings for his own thoughts). Of note is that James's spiritual self, in part, reflected his metaphysical leanings, including his interest in religion and the afterlife. As we will see in Part II, both *spiritual* and *scientific* ideas comfortably cohabitated in nineteenth-century biologists called "vitalists" who believed that life was due to some immaterial substance, sometimes called *élan vital* (vital impulse).

By the 1920s, investigations into the conscious self and other aspects of the mind had been put aside—in fact, banned—while the behaviorists reigned over the next several decades. Behaviorists objected to the foundational role that the conscious mind played in psychology because it was unobservable by anyone besides the introspecting person. Psychologists, the behaviorists said, should not be both the experimenter and the entity experimented on. Similarly, they objected to Sigmund Freud's views about the unconscious mind, which was unobservable even to the introspecting person. From the behaviorist perspective, psychology should, like other areas of science, be based on observable and measurable variables, which in the science of psychology were stimuli and responses.

The emergence of cognitive science in the middle of the twentieth century weakened the behaviorists' hold on psychology, and in fact brought the concept of mind back to the field. But it was not the same self-conscious mind that the forefathers of psychology had been concerned with. The cognitive mind was, instead, portrayed as an information processing system, rather than a system that makes subjective experiences. It was in a sense unconscious, but not in the deep Freudian sense. The unconscious aspects of cognition were mostly framed as processes that support mind and behavior behind the scenes without explicit conscious content—processes that create representations by manipulating symbols. Cognitive science was not anti-conscious, but it was disinclined to investigate it, since so much could be explained without making recourse to consciousness. Over time, as behaviorism receded in psychology, consciousness, including the conscious self, slowly began to make its way back.

These days, the seventeenth-century Western philosophical fusion of self and consciousness, which objectified the self as a natural entity of the physical world, is alive and well. This is why we think of who we are as an individual in terms of the word *self.* It is why we treat our self as an entity with properties and call on this entity to explain our thoughts and actions. It is also why the self is a large and thriving research area that spans philosophy, psychology, psychiatry, neuroscience, biology, and many other fields.

"Self"-Criticism

So what's the problem? For starters, there is little agreement about what the self is. For some the self is just as Locke said: a consciousness. For others there are hidden or unconscious aspects of the self. There are, in fact, many theories of the self, and many different ways to talk about it and measure it. But for something to exist in nature as a real thing that can be documented scientifically, it has to exist independent of both the way it is measured and the language that is used to describe it. We don't study the self. We study ideas about the self. Why? Because the self is itself just an idea.

I am hardly the first to critique the idea of the self as a real entity. Shortly after Locke's time, the philosopher David Hume referred to the self as "the elusive 'I' that shows an alarming tendency to disappear when we try to introspect it." With the rise of scientific psychology, the value of "self" as a topic was heatedly debated. More recently, but in a similar vein, the philosopher Daniel Dennett and others have described the self as illusory. Some have noted that discussions about the self are really discussions about other things; the self is a gratuitous notion. I think Thomas Metzinger truly hit the nail on its head when he said that "nobody ever had or was a self." A similar perspective has recently been offered by the psychologist Zoltán Dienes.

Another obvious problem with the traditional notion of the self is that it remains a narrowly focused conception promoted by European philosophers and imported into modern psychology and neuroscience. Cultural traditions that treat the self and other aspects of human existence differently have received less attention.

Efforts have been made to get around the various critiques. The philosopher Shaun Gallagher pointed out that a common strategy is to avoid directly focusing on the self by modifying it with other terms. Examples include the conscious, unconscious, core, autobiographical, narrative, existential, social, embodied, and synaptic self.

Similarly, the psychologist Ulric Neisser suggested a focus on *aspects* of the self, as opposed to the self itself. And the philosopher Paul Thagard has listed more than eighty words related to self, dubbing these *self-phenomena*. These are useful ways to talk about findings that have come from self-research, since the *aspects* or *phenomena* capture important things related to how we think, imagine, remember, hope, regret, and believe about who we are. But some go further and use such findings about the self to generate a bottom-up conception of what the self actually is. For them, the self can be thought of as an emergent property.

Emergent Selves

As explained in a 2017 Technology Report by the Aspen Institute, "Emergence is what happens when a multitude of little things—neurons, bacteria, people—exhibit properties beyond the ability of any individual." Ant colonies, cities, bodies, brains, and life itself are other common instances. But emergence is not limited to living things. A mountain, for example, is a description of a state of nature that emerges from interactions of soil, minerals, moisture, decaying organic material, and weather.

Two versions of emergence are distinguished. Weak emergence is descriptive—in this version, the *self* is simply a word that summarizes the combined effect of all the underlying phenomena that have been discovered about the self. In strong emergence, by contrast, the self is given agency, the power to do things. Weak emergence is commonplace, since we know that complex things emerge from combinations of less complex ones. In strong emergence, which is quite controversial, the complex outcome is not an obvious result of the parts acting together.

Many self-models adopt strong emergence, at least implicitly. For example, Thagard suggested that a *self system,* capable of doing things cognitively and behaviorally, emerges from self-phenomena. Andrzej Nowak's *society of self* similarly proposes that low-level psychological processes contribute to an emergent *cognitive structure* that is self-organizing and dynamic, and that is a controlling agent for lower-level psychological processes. Gallagher refers to the self as emerging from dynamic, organized patterns of brain activity that affect who one is and how one acts moment to moment. Eric LaRock and his colleagues also explicitly argue for a strong-emergence view of the self, writing that "what emerges are not merely mental properties in specialized, distributed neural areas, but also a new, irreducibly singular entity (i.e., an emergent self) that functions in a recurrent (or top-down) manner to integrate its mental properties and to rewire its brain." In strong-emergence views, the self exists over and above the underlying phenomena that constitute it. I admit that I too went in this direction in *Deep History* but now see things differently.

James succinctly characterized why he thought the self is so hard to conceptualize, saying that it is "partly known and partly knower, partly object and partly subject." Well over a century after James's death, the situation has not improved, and may actually be worse. As Gallagher has pointed out, despite the field's current comfort in talking about the self, what is said is usually controversial. In *Being You: A New Science of Consciousness,* Anil Seth writes that although it may seem that your self is the thing that perceives the world, your self is instead just another perception. The philosopher Owen Flanagan put it this way: "Just because 'self' is in our vocabulary does not mean that it has any explanatory role in a science of the mind."

For my part, I believe that the mystery of what the self is would disappear if the self were treated as a descriptive abstraction about an individual, rather than as an entity within the individual that does things for the rest of that individual. Since this does not seem to be the way that research and theorizing about the self are going, a reboot may be required. As we will see next, similar problems plague the self's younger sibling, personality.

3

The Personality Contest

The late nineteenth century was a period of intense activity related to the mind. Experimental psychology, psychiatry, and brain research were all emerging. It was also a time when the notion of *personality* began to offer an alternative to *self* as a way to think about what a human being is.

As a term, *personality* is derived from the Greek *persona,* which referred to a theatrical mask that expressed traits other than those of the individual wearing the mask (recall that the Greeks did not have a word for self per se, but did talk about *persons*). Later, *person* would come to have spiritual, legal, and ethical implications. But it was not until the late nineteenth century that the psychological meaning that personality has today took hold, when the French philosopher Charles Renouvier used the term to refer to the *embodied* and *empirically knowable* individual. French physicians around this time began to treat one's personality as something prone to illness and disease, and consequently, like any other medical problem, potentially treatable. An 1885 book introduced the notion of *alternating personality,* which later came to be referred to as *multiple personality disorder,* and more recently, *dissociative identity disorder.*

The medicalization of personality in the nineteenth century can be seen as an effort to conceive of what people are in a more grounded way, bodily and behaviorally, rather than only in the Lockean terms of an ephemeral conscious self that could be known only privately by that same self. The corporeal body thereby became a key part of personhood. But as we will see, as with self, there have been many different theories of what personality is. And although the term personality is sometimes used as a broader conception that subsumes the self, it is also sometimes used interchangeably with self.

Freud and the Gang

It was common in the late nineteenth century to distinguish between neuroses and psychoses. Neuroses, such as anxiety and hysteria, were thought to be nervous-system problems (hence the root "neuro"). Sigmund Freud, as a young physician and aspiring brain scientist, developed an interest in the neural basis of neuroses, but soon realized that the science of the brain was too primitive to be useful. His encounters with patients led him to redefine neuroses as psychological problems, and he developed a psychological approach to their treatment called psychoanalysis that popularized the *unconscious,* a notion that he borrowed from earlier philosophers, especially German philosophers like Leibnitz, Kant, and Herbart.

When referring to the mind, Freud sometimes used the German expression *Das Ich* (the I), but rejected the Lockean notion that a continuous conscious self over time is the core of a person. The conscious self, or ego, was just one part of the mind, with the id and superego being the other two. For Freud, unconscious aspects of mind and behavior (instinctual impulses and needs of the id) were more important than consciousness; he derided consciousness as the "tip of the iceberg." The unconscious root of neuroses, he believed, typically appeared in early childhood, often in relation to experiences surrounding sexual development. Freud's views were, in general, more compatible with the idea of a troubled person than simply a troubled self (ego). That is why his psychoanalytic model of the mind is considered a kind of personality theory.

Psychoanalysts who followed Freud added their own stamp to psychoanalysis, breaking with him in various ways. Some dropped the emphasis on sexual impulses, others minimized instincts in general, and others deemphasized the unconscious. The psychoanalytic view of personality was becoming more eclectic, which led to the self having a more significant role.

Carl Jung, for example, believed that Freud's idea of the unconscious as a repository of negative states was too limited. Jung stressed the idea of a collective unconscious, which he proposed housed archetypical symbols shared by all members of the human species, as opposed to the kind of personal unconscious that Freud emphasized. For Jung, one's self differentiates into an integrated person out of random processes in the collective unconscious, at least when things go well.

Other Freudian analysts started a movement called *ego psychology.* For example, Harry Stack Sullivan proposed a *self-system,* a set of supervisory processes that protect one from anxiety by sanctioning some behaviors and discouraging others, and that serve as the gateway to consciousness (an idea reminiscent of Bishop Butler's self-monitoring system, mentioned earlier). Sullivan also boldly proposed that personality is an illusion, a hypothetical construct that is meaningful only in terms of social, interpersonal relations.

Erik Erikson, another psychoanalyst, rejected Freud's emphasis on rigid instincts, arguing instead for the importance of the ego in learning and of *plasticity* in the context of culture. He viewed one's ego identity as being anchored in cultural identity, and coined the term *identity crisis.* With Erikson and other post-Freud Freudians, the distinction between self and personality dwindled.

Personality as Behavior

One of the more common views of personality is the way that an individual seems to others who observe how they act in the world. The behaviorist movement in psychology was thus, in some ways, a perfect fit with this version of personality theory.

Behaviorism arose in the early twentieth century in part as a reaction to a previous over-dependence on unobservable conscious and unconscious mental states to explain human and animal behavior. B. F. Skinner, a leading figure during much of the behaviorist era, had little interest in formal theories, preferring a descriptive approach that accounted for the relation between a behavior and its consequences. For example, when a behavior has a useful consequence, the likelihood that it will be repeated in similar situations in the future increases. Skinner believed that learning was a universal process that could be studied the same way in animals and humans, and he did most of his work on pigeons. Toward the end of his career, he began applying these findings to help people modify undesirable behaviors. His novel, *Walden Two,* brought his major ideas into the public eye, but not completely favorably. Behaviorism was viewed as a cold, impersonal approach to psychology. Even so, other behaviorists also went in the therapeutic direction, resulting in what came to be called *behavior therapy.*

Humanists

The humanist movement in psychology, which emerged in the 1950s, was meant as a course correction to both Freud and Skinner—Freud for his dark side (his focus on negative emotions like anxiety) and Skinner for his cold objectivism (his rejection of subjective experience) and reductionism (his explanation of complex human behaviors using studies of animals).

Abraham Maslow advanced a humanist theory of personality that emphasized positive values related to human societal goals. He proposed that from birth we have a *will* toward growth and health; that is, toward self-actualization. Neurosis was seen as a thwarting of one's potential by the sickness of society. Maslow took the radical step of finding and studying people who he described as *self-actualizers,* people who had *peak experiences,* assuming this might reveal how to help improve the lives of non-actualizers.

Another humanist, Carl Rogers, initiated a *person-centered theory* that emphasized the *inner life* of individuals and their personal perspective. For him the world was a *subjective reality* consisting of both conscious (symbolized) and unconscious experience. Particularly important was the congruence between the internal self and the self as perceived in subjective reality. Anxiety was seen as a result of the incongruence between one's subjectively perceived and ideal self.

The Maslow-Rogers tradition lives on today in the extremely popular *positive psychology* movement.

Traits and Temperament

Personality is often associated with enduring, entrenched traits that characterize the way people are. These are said to be difficult to change because they are mostly unconsciously expressed, and based either on inherited genes or on rigid habits. As Theodore Millon put it, "Personality narrows one's repertoire to particular behavioral strategies that become prepotent and typify the individual's distinctive ways of dealing with others and one's self."

No other approach to personality has generated more research than trait theory. It has ancient roots. The Greek philosopher Hippocrates and the Roman physician Galen both described personality types based on body humors (what we call hormones today).

Modern trait theory began in the late nineteenth century through the work of Francis Galton. He used questionnaires to collect data about people's backgrounds and behavior, and applied statistical analyses to make sense of the results. He also introduced the use of common-language words about mind and behavior to label his bottom-up statistical findings about personality. Greatly influenced by Darwin, Galton believed that human physical and mental capacities were inherited traits, and was among those who embraced eugenics to improve the quality of human genes by excluding from society those thought to be inferior.

Despite the stain of eugenics, the trait approach was reinvigorated in the twentieth century by Gordon Allport. Early in his career Allport visited Freud and left feeling that to understand personality he needed to focus on factors more obvious than those hidden in the depths of the unconscious. Like Galton, he took a bottom-up approach, compiling a list of more than four thousand words related to what a person is. Raymond Cattell also applied statistical analyses of people's self-reported responses to questionnaires. He whittled down Allport's much longer list to just sixteen words.

Currently, the so-called big five model is the most widely accepted trait scheme. It proposes five trait groupings: openness, conscientiousness, extraversion, agreeableness, and neuroticism. Within these are higher-order *clusters* and lower-order *facets* of personality (for example, in the extraversion cluster, warmth and gregariousness are facets). Both genes and experience are thought to affect the big-five traits.

Trait theory has its critics. For example, Henry Murray, who dubbed his motivational theory of personality *personology,* chided trait theorists for reducing personality to statistical events, asking, where is the individual, the person, in trait research?

A popular contemporary variant on trait theory emphasizes stable temperaments based on genetic differences. Jerome Kagan, for example, underscored how an *inhibited* temperament, starting in early life, predisposes one later to develop anxiety. C. Robert Cloninger's three-temperament domain theory is also popular and has been related to neural and genetic correlates.

Situational Personalities

In the 1970s, the behaviorist Albert Bandura introduced factors such as *expectation* and *observational learning* that tilted toward the newly emerging social and cognitive approach to psychology. A colleague of Bandura's, Walter Mischel, took the social cognitive approach further. He was a staunch critic of trait theory, arguing that behavior is shaped by the situation one is in. Specifically, he maintained that while behavior can be very consistent when situations are similar, it can also be quite different across different kinds of situations. The reason for this, Mischel said, is that how we act in a given situation reflects cognitive appraisals of our personal relationship to the situation. For example, one may be introverted in work situations, and extroverted among family and friends. In other words, we have different personas, an idea that goes back to Cicero and Locke.

A related view is social construction theory, which has its roots in anthropology and other social sciences. It posits that one's personality is shaped by social and cultural contexts, and that self and personality cannot be understood without these factors. The social cognitive point of view is somewhat reminiscent of the views of analysts Sullivan and Erikson.

Embodiment

Philosophers interested in personality have traditionally followed Locke's position, treating personhood as essentially consisting of conscious states about one's self. More recently, some philosophers, like the early personality theorists, have emphasized the difference between *person* and *self.* Peter Strawson, for example, proposed that persons have both conscious selves and corporeal (body) characteristics, making *person* a broader conception than *self*, or at least broader than the conscious self.

There is, in fact, a growing movement called *embodied cognition* that treats the mind as being associated not just with the brain but also with the rest of the body, and, for some, even the outer environment (so-called extended cognition). Personality was embodied from the get-go (recall its start in France as an embodied condition that was an alternative to the conscious self), but in the cognitive embodiment movement, even the self is, in part, embodied.

For example, the philosopher Mark Johnson asked the question, "What do we mean by 'the body'?" It's a good question, since *body* implies the non-brain components of the body. Johnson asked his question to set the stage for an argument that the self can (or, for him, should) be understood as an embodied form of cognition—that self-cognition extends into the non-neural body.

Thomas Metzinger proposed a hierarchy of self-embodiment in which the conscious phenomenal self depends on one's brain and body, and Shaun Gallagher introduced the idea of an embodied minimal self, which is a more primitive state than the self-conscious self. The neuroscientist Antonio Damasio has long embraced the idea of bodily components of self, in referring to a *proto-self*, and Jaak Panksepp, another neuroscientist, calls this the *core self*, but both also propose a higher cognitive conscious self that is self-aware.

In my 2001 book *Synaptic Self*, I also suggested that the self has conscious and non-conscious aspects; by the latter I meant implicit processes that define us in some way, but that do not reach conscious awareness. Primitive embodied aspects of self, such as those discussed by Damasio and Panksepp, were, in my model, part of the implicit self. I likened these unconscious aspects of the self to personality.

Two entwined issues in this discussion need to be pulled apart. The first is the relation of self to the body. I fully accept that fundamental phenomena related to the proto / core / minimal / embodied-self idea capture basic, relatively primitive, aspects of who we are. But I do not think these phenomena need to be burdened with the *self* moniker, and all the conceptual baggage that comes with it. The second issue is what is meant by embodiment. In the early days of cognitive science, cognition was all about creating representations of the external environment. Few today would question that cognitive functions of the brain monitor body signals and make sense of the internal environment of the body much the same way that they use visual or auditory signals to monitor and make sense of the external world. While not in your body per se, cognitive functions of the brain depend on your body to say alive. By the same token, the survival of your body depends on your brain's neural, including its cognitive, functions.

Persons in Flux

What is a person? In the simplest sense, the answer is an individual human being. One's biological, especially genetic, makeup differs from that of all other persons. But individuals do not typically live in isolation. Family and community contribute, as does one's culture. Indeed, biology is no longer considered the sole basis for sexual identity, long thought to be a fixed biological trait.

The question of who, or what, is a person sometimes becomes a legal issue. Does a person's life begin when the egg is fertilized, at some point during the fetal stage, at birth, or upon psychological maturity? At what point in the course of dementia does one's legal status as a person cease? Peter Singer and some other philosophers make a distinction between humans and persons. Singer argues that there can be humans who are not persons—such as people with severe dementia—and that some animals can be persons. In the age of artificial intelligence, some even predict that as cyborgs become more sophisticated, they too will deserve to be considered persons with rights.

Personality / Self Redux

Personality theory offered a broader perspective on what a person is than self-theory could. But as the notion of the self began to make its way into these theories, and even became the main focus of some, differences between self and personality began to diminish. And when the self became embodied, and even unconscious, self and personality sometimes became theoretically indistinguishable.

Earlier I wrote that the self is not an entity within us that does things. Personality theorist Henry Murray made a similar point, noting that personality is an abstraction that exists more in the minds of theorists than in the minds and bodies of people. This may be why there is so little agreement in academia about what self and personality are. But conceptions of self and personality serve important personal, social, and cultural purposes. It is, therefore, important that we get this right, even if it means jettisoning the popular strong-emergence view of self and personality, which considers those to be entities with agency that account for who we are and why we do what we do.

4

It's Only Words

We scientists take pride in being methodologically rigorous. We are very picky about how we design our research projects, collect data, and statistically analyze the findings. But we can be less meticulous about how we interpret what it all means. This shortcoming was expressed well in 1995 by the personality psychologist Jack Block at the end of his career:

> I urge the field of personality psychology to resolutely confront its severe, even crippling, terminological problems. Many of the difficulties that beset assessment derive from the hasty, hazy, lazy use of language. Psychologists have tended to be sloppy with words. We need to become more intimate with their meanings, denotatively and connotatively, because summary labels and shorthand chosen quickly will control—often in unrecognized ways—the way we think subsequently. In part, this problem is inevitable, but we can do much better.

Block's plea to his colleagues could be directed to many areas of psychology and neuroscience, including, of course, studies on the self. Indeed, I believe that, in no small part, confusion about how to characterize, and understand, what and who we are is terminological. Some might say, "Oh, it's just semantics," as if that means it's not important. But it is. Words matter. They dictate meaning and underlie our conceptions. This is especially a problem with the words used in psychology.

The Language of Psychology

Many have noted that psychology is different from the other sciences because it is more dependent on commonsense wisdom handed down through the ages and incorporated into everyday language. For example, in their 1959 book *The Language of Psychology,* George Mandler and William Kessen noted: "The common language is full of quasi-psychological assertions. . . . The fact that man studies himself and that he has archaic notions which persist in the daily behavior of all men puts a major stumbling block in the path of scientific psychology."

It doesn't matter much if physicists or biologists are lax with their use of common language terms. No one really believes that words like *string* in string theory or *hedgehog* in genetics are accurate descriptions of what they refer to (particles and genes, respectively). They are simply catchy terms that have some metaphorical relation to the real deal. But when we use a mental-state word to name some functional process or circuit in the brain, the mental state is often assumed to be what the function is, or what the circuit does. In other words, the function or circuit inherits the mental-state qualities, and the conceptual baggage, implied by the name.

Kurt Danziger, in his insightful book *Naming the Mind,* noted that the reason we talk the way we do today about psychological processes is often because long-ago philosophers and scientists with limited or no knowledge of the brain assumed that if behavioral responses and conscious experiences in humans occur at the same time, the conscious experience must have caused the behavior.

When ideas are discussed the same way generation after generation, they come to be so natural, so intuitive, so seemingly true, that they go unquestioned. Although scientists are taught from day one not to confuse correlation and causation, they too can mistake the intuitive for the factual.

An example near and dear to my heart, and one that I will discuss later in more detail, is the notion that the feeling of fear is what causes us to freeze or flee in the presence of danger. This bit of folk wisdom, or what is known as *folk psychology,* does not reflect the way things work in the brain, even though it is the accepted view of researchers and lay people. Don't get me wrong. I am not against folk psychology. There is a place for folk-psychological mental-state terms. To reiterate an earlier point, they can and should be used to describe mental states like fear and the circuits underlying such conscious states. But they should not be used to describe circuits that control behavior non-consciously.

Explaining by Naming

Another problem that scientists have with vernacular words is reification. Typically, when you choose to study something scientifically, you do so assuming that the thing in question exists as a natural phenomenon in the physical world the way a rock, tree, building, or animal does. If the thing you want to study is not part of the physical world, you can't measure it. The measurement process is what allows you to validate the actual existence of the thing referred to by the name. But sometimes naming gets mixed up with explaining.

In 1620, Francis Bacon wrote, "Scientists should be vigilant . . . and especially should guard against tacitly granting reality to concepts simply because we have words for them." In short, when we apply names to philosophical ideas or scientific results, we run the risk of reifying them, endowing them with the properties implied by the name we give them.

Consider psychological *essences,* hypothetical concepts or processes that are assumed to exist as real entities. As Cameron Brick and colleagues put it, "Psychological concepts (e.g., intelligence; attention) are easily assumed to represent objective, definable categories with an underlying essence. Like the 'vital forces' previously thought to animate life, these assumed essences can create an illusion of understanding." Essences are intuitively appealing, in part because they have been handed down culturally over generations. But as thoughtful psychologists have pointed out, such constructs typically lead to theoretical dead ends.

The philosopher Kenneth Schaffner, who specializes in philosophical issues related to psychiatry, suggests that mental disorder constructs (schizophrenia, anxiety, depression) should not be thought of as set in stone, but instead as contingent and changeable, or even dismissible, as knowledge changes. He proposes that we treat disorders not as entities, but as collections of similar prototypes.

Intelligence is an excellent example of how naming and measuring lead to reification. In the late nineteenth century, Francis Galton proposed that intelligence, like height or weight, is something you have and that it varies in degree—so it can be measured. Tests were designed, resulting in an intelligence quotient (IQ). Intelligence came to be thought of as an inherited factor, and was sometimes used, as Galton did, to promote eugenics and other racist ideas. Eventually, IQ tests, like many standardized tests, were shown to be culturally biased. There is no measurable entity in a person that constitutes their intelligence. Some have suggested that intelligence is nothing more than what an intelligence test measures, which is, in the end, a decision based on someone's ideas about what intelligence is. If intelligence does not exist independent of its measurements and the way it is talked about, it is not a thing. It is an abstraction, one that reflects a variety of cognitive, emotional, social, and other skills. One acts "intelligently" because one's skills combine in a certain way that is useful for solving some problem or problems, not because the skills are products of the amount of intelligence they possess. It's important to understand the variability of these skills across individuals and their effects on mental and behavioral processes, but not because they collectively implicate an actual entity in the brain called intelligence.

Recent studies have shown that dynamic, temporary coalitions of neural networks emerge when people perform tasks related to capacities measured by intelligence or personality tests. The way to think of these networks is not in terms of entities in the brain that make one *intelligent* or give one a *personality.* Instead, the networks simply reflect the activity of underlying processes that are engaged when one performs the tasks in question. This is weak rather than strong emergence at work.

Agent Provocateur

Terms like self and personality have played an important role in helping to organize facts about certain aspects of mind and behavior. The problem occurs when we objectify self or personality as entities with agency, things within us that control what we do.

When we assign agency to an entity within us, it allows us to credit our self for our achievements and blame our self for our failures. But suppose one's failure to be a better pianist is due to tiny defects in the bones in one's hands, or in the nerves that move the muscles that control the bones. Or suppose one's success as a chef is due to an unusually acute sense of smell. Consider, too, that if one works really hard at something, the self is usually credited, and if one does not put as much effort into a task, the self is typically blamed. But suppose the hard worker, because of personal circumstances, had the luxury of spending more time on the task, whereas the other did not. These are traits, characteristics of the person. But the person as a whole does not deserve the lion's share of the blame or the credit for these outcomes resulting from biological, circumstantial, or social factors.

This is not to say that one never personally deserves credit or blame. The point is instead that there is no singular self-agent that is charge of what we do. As I will discuss later, there are many levels of behavioral control in the brain, and most are *not* under the direction of one's self-conscious mind. And for those that are under conscious control, we don't need to refer to a self to explain that control.

Our most common uses of *self* involve personal (I, me, him, her, they, it) or reflective (myself, himself, herself, themselves) pronouns that occur with adjectives (self-assured, self-aware, self-inflicted, self-hating) or nouns (conscious self, narrative self, social self, spiritual self). Each of these uses is nothing more than a description of an individual, or of features of an individual. None involves an entity that is in control of that individual. If I say, "I ate the cake myself," I don't mean that some entity called Myself is the thing that ate the cake. I am just pointing out that the person I refer to as *I,* as opposed to some other person, ate the cake. The self is just a nickname, a description, we apply to systems in us that do things.

The self, in short, boils down to the cognitive capacities that monitor your body and mental states, and that use memories of who you are in relation to your present needs, expectations, capacities, and limitations to make predictions about how to act. Since, by definition, it is you (your brain and body) that is doing all this, calling on a self to explain what is being done is redundant, and hence unnecessary. You are your self, and your self is you. There is no other entity to find inside you. It's just you in there because your self is simply an idea you construct about who you are. The pioneering cognitive psychologist Jerome Bruner nailed it:

> Self is a perpetually rewritten story. What we remember from the past is what is necessary to keep that story satisfactorily well formed. When new circumstances make the maintenance of that well-formedness sufficiently difficult, we undergo turning points that clarify or "debug" the narrative in an effort to achieve clearer meaning.

The wedge I am driving here is between the poorly understood, over-aggrandized, notion of the self as an agent in charge, and the concrete, mundane psychological and neural processes that do the stuff the self is said to do. Thinking of these processes as being under the control of the self is not only unnecessary and redundant, but also scientifically counterproductive.

I don't mean to imply that we lack conscious agency, so-called free will. Instead, I am saying that much of our behavior control occurs non-consciously, including a good deal of behavioral control misattributed to a conscious self. We desperately need a clearer understanding of which kinds of behaviors are controlled consciously versus non-consciously. I will argue that viewing behavioral control in terms of our realms of existence provides that understanding.

Personalized Bodies and Brains

Our bodies and brains are personalized. No one else has your exact genetic makeup nor the exact epigenetic events that make your body, including your brain, unique. And no one else has your psychological uniqueness, or what Carl Rogers referred to as one's *personal perspective.* This *you-ness* is obviously shaped by "your past"; by memories you have formed living your life. Distinguishing what is yours from what is not yours with a word like your personality adds nothing. Everything in your brain and body is already yours.

A personalized brain and body, one might say, has a personality. That's fine, if by personality you mean a description of you. But once you make personality something more, an entity in charge of what you do, you are reifying an essence that does not otherwise exist.

5

A Path Forward

When ideas become entrenched in tradition, and seem so familiar and well established that they must be true, it is sometimes useful to put them aside and ask whether there might be another way to account for the relevant phenomena. As should be abundantly clear, I think this is where we are with self and personality.

I propose that the best way to understand what, and who, we are is not from the lofty heights of philosophical reasoning, or from even data-driven, bottom-up scientific views about high-level constructs like self and personality. Instead, I believe that the path forward is through understanding the kind of living beings we are.

Four Hierarchically Integrated Ways of Existing

My core conception is that, at the present stage of evolutionary history, there are four basic ways of existing on Earth: biological, neurobiological, cognitive, and conscious. These *realms of existence* define what kind of life a living thing lives.

Every living thing that has ever existed has existed biologically. All biological beings, all organisms, possess basic life-sustaining processes such as metabolism, as well as the species-sustaining capacity for replication. Most living things exist only biologically.

But some biological beings possess a nervous system to assist with mere biological existence. On this score, members of the animal kingdom are unique among biological beings. Animals, and animals alone, possess nervous systems, and hence are the only *neurobiological beings.* Nervous systems make possible greater speed and more precise body control than exists in any other kind of organism.

In some animals, mere neurobiological existence is supplemented with the ability to create internal representations of the environment and use these to construct mental models that make predictions about the world—all of which endow those animals with greater flexibility in responding behaviorally to challenges and opportunities in life. And some, but not all, cognitive creatures also exist consciously. Deciding which creatures are cognitive and conscious beings is a contentious topic that I'll address later.

Of course, just because two groups of animals exist in the same general ways does not mean that they exist in the exact same ways. Worms, bees, fish, frogs, lizards, birds, cats, dolphins, monkeys, and people are all neurobiological beings, but they differ considerably in how their brains and bodies enable their behavioral interactions with their surroundings. Similarly, evidence for cognition exists for vertebrates and possibly some invertebrates, but the varieties and degrees of complexity of what exists differ considerably across groups and even within groups. The same holds for conscious existence—two different kinds of animals may both be conscious, but that does not mean they are conscious in the same way. Because words like cognition and consciousness were invented by humans to describe what goes on in the human brain, these terms should be used judiciously when discussing other animals.

Our realms are hierarchically related. The conscious realm, according to the point of view I am developing here, is enabled by the cognitive realm, which depends on the neurobiological realm, which relies on the biological realm. Each realm anatomically per-

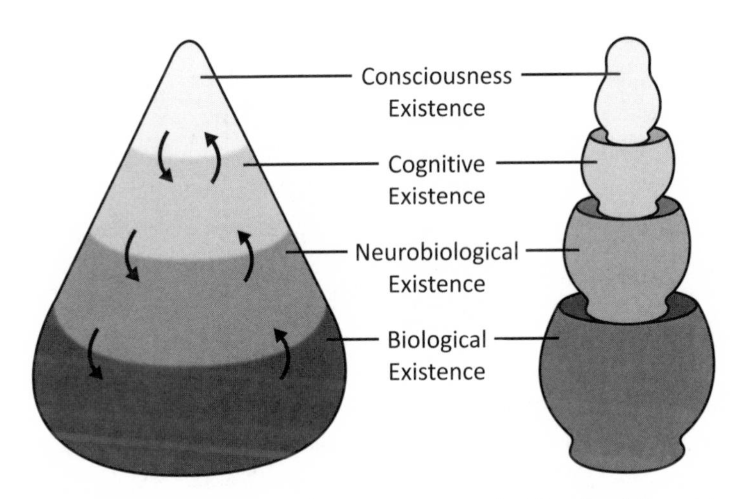

Figure 5.1. Entwined realms of existence versus stacked-Russian-doll model

meates and physiologically enables the level above it, and at the same time, the survival potential of the level below is enhanced by the one above.

In a sense, our realms of existence resemble components of a Russian doll. But unlike a Russian doll, in which one part nests within, and stacks on top of, another, our component realms are conjoined and interdependent.

The Body as a Repository of Evolutionary History

The realms of existence of a particular individual animal are constituted as living globs of entwined protoplasm that reflect just one of the many, many outcomes that have resulted over billions of years of evolution. Through natural selection, the body plans of existing species were adapted and modified in ways that provided additional advantages in their present, but ever-changing, environment. Of course, some of these modifications also eventually led to the emergence of new species.

The evolutionary points where radical, as opposed to incremental, changes took place are called *major transitions.* Tyler Volk, my long-time colleague at NYU, and co-founder with me of the rock band the Amygdaloids, described evolutionary transitions as stages of relative continuity that are ratcheted one to the next by changes of state. Volk referred to the resulting state changes as *dynamic realms,* which inspired my notion of *realms of existence.*

The primordial, biological, realm has existed for roughly 3.7 billion years. It sustains life, and it is a preexisting condition for the persistence over time of any kind of living thing, whether a single-cell microbe, such as a bacterium, or a complex multicellular organism, such as a plant or animal. The neurobiological realm arrived between six and seven hundred million years ago, when animals invented nervous systems. Nervous systems evolved as extensions of, and supplements to, the biological way of life, allowing coordination of spatially separated internal body functions, and more efficient control of behaviors directed toward the external environment. These basic neural processes were then the template from which some animals evolved the ability to meet their survival needs by using internal representations and mental models to flexibly choose among hypothetical options for action without having to risk the consequences of actually testing the options in the environment. Refinements of these capacities became the cognitive foundation of the kind of mental-state consciousness we humans possess, which allows each human to consciously know that we are a living thing with a personal past and some undetermined future. This does not mean that all other animals lack consciousness. It just means that those that are conscious are not conscious in the same way we are, just as they do not walk or communicate the way we do. By definition, different species have different bodies, and consciousness, when it is present, is part of the body and must vary between species like any other body function.

All four realms are present throughout our adult life, contributing to our most basic experiences of pain and pleasure, hunger and thirst, disgust and lust, as well as more complex feelings of love and hate, compassion and hope, despair and ecstasy. *Everything* about one as an individual human being, biologically and psychologically, is subsumed within the entwined, hierarchical organization of our realms of existence.

As an illustration of the interdependence of the four realms, consider what underlies a human conscious experience. Every such event is preceded and enabled by non-conscious, or more precisely, preconscious, cognitive processing. And both consciousness and cognition depend on underlying neural activity. Neural activity, in turn, requires that brain cells use their metabolic capabilities to make energy. Metabolism requires that the products of digestion (nutrition) and respiration (oxygen) be pumped through the circulatory system to the brain by the heart. But the delivery of nutrition into the body requires food-seeking behaviors controlled by the brain, and these behaviors, in turn, depend on metabolism in the body. In animals like us, often cognition, and in some instances, consciousness, is involved in planning what to eat, where to get it, and when to consume it. Each of these processes engages each of our hierarchical realms of existence.

You might be thinking that this way of describing who we are sounds like I am embracing the notion of embodiment. But for me, the mind and brain are *not* embodied. Instead, the body is "embrained." Okay, that's a bad use of language, but it makes the point, which is that the body is represented in the brain via the capacities of the neurobiological realm. This neurobiological representation of the body is then re-represented in the mind via the processes of the cognitive realm, and some of these cognitive representations of the body are (or can be) subjectively experienced via the conscious realm. That's how cognition and the body interact.

A case could be made that social and cultural factors constitute additional realms of existence. But that would reflect a misunderstanding of what I mean by a realm of existence. Society and culture are things that people do, or have done, with their four realms—they are ways of living, behaving, thinking, and feeling, and in general, being, rather than additional realms.

Ontological Modes

Realms of existence are analogous to what philosophers call *ontological modes.* For example, Aristotle claimed that human beings have vegetative, animative, and rational ontological modes, or simply, ways of existing. There have been many such proposals over the centuries by philosophers, psychologists, neuroscientists, and biologists. This does not mean I am just parroting what came before. Perhaps the closest position to mine is Thomas Metzinger's *hierarchy of self,* mentioned in Chapter 3. Other positions related to mine are from Daniel Dennett and Simona Ginsburg and Eva Jablonka. These authors focus on the features that distinguish organisms throughout evolutionary history. I also did this in my book *The Deep History of Ourselves,* and to some extent, do that here. But my focus in this book is less about cataloguing which realms are present in different animals, and more about how our four realms, especially the three that depend on having a nervous system, make us what and who we are as humans.

No other animal has a human brain, and to the extent that other animals have realms that overlap with ours, their realms will function differently than ours—for example, monkeys and humans have similar cognitive realms, but monkey cognition differs from human cognition. And while every human being, barring congenital or other disease conditions that alter brain function, has all four realms instantiated in the human way, one's realms, like one's other body features, are also instantiated in a unique, individual way—each person's realms differ from the realms of every other individual. Differences in how one's body, including the brain, have been affected by one's biological (genetic) makeup, mentioned earlier, are one reason for this; the unique life experiences that have been stored as memories are another.

Momentary Realms Form Ensembles of Being

One's realms of existence, in the moment, collectively constitute an *ensemble of being.* You might be asking, Why is an *ensemble of being* arising from *realms of existence* any less abstract than self or personality? Actually, it's not. Like self and personality, an *ensemble of being* is merely a descriptive concept. But unlike what has often happened with the ideas of self and personality, I am making no claim that an emergent, momentary ensemble of being is an actual entity that does things. It is not a system that organizes or controls who we are or what we do. It has no awareness or agency. It is not, itself, the core of who we are. It is simply a description of the integrated activity of the separate but entwined activities of the biological, neurobiological, cognitive, and conscious realms. And because one's realms are dynamic—continuously changing—the ensemble of being also changes continuously.

Clearly, neuroscientists could find neural signatures that characterize such momentary ensembles of being, much the way that they can already document the temporary coalitions of neural activity that occur when people perform tasks related to intelligence or personality tests. Presumably, with repeated measurements, individual neural signatures (*being prints*) could be found. But just as neural patterns that relate to intelligence and personality do not make one intelligent or give one a personality, the neural patterns related to ensembles of being would reflect, but would not make you, who you are. They would be correlated with, but would not cause you to be, who you are.

Since one's momentary ensemble of being does nothing, it needs no explanation as a cause of anything. It is just a description, an abstraction, that emerges when we put together current knowledge about (1) the biological requirements of life across the evolutionary history of organisms, (2) our neurobiological understanding of the brain and its control over body physiology and behavior, (3) our neurobiological and psychological understanding of cognition, and (4) our neurobiological and psychological understanding of consciousness.

Why Bother?

If ensembles of being don't do anything, why should we care about them? The various mechanisms and processes that constitute each realm are a distinct category of organismic activity. They keep us alive, and allow us to behave, know, feel, and imagine possibilities, and, in general, make us what and who we are. In short, we care about ensembles of being because they reflect the composite activities occurring moment-to-moment in the realms.

By providing a novel view of what underlies the individual, personal nature of human existence, this construct—ensembles of being—fosters new ideas about what it means for individuals to not just survive, but also thrive. For example, understanding how our intricately entwined realms span biological, neurobiological, cognitive, and conscious ways of existing eliminates the awkward and tired distinction between physical and mental, between biological and psychological, well-*being*. More generally, it forces us to see that achieving and sustaining well-being cannot be done piecemeal, since problems in one realm affect the vitality of others. We thrive as individuals when our realms are integrated as one; we suffer when they are not.

PART II

THE BIOLOGICAL REALM

6

The Secret of Life

The biological realm encompasses the entirety of life, from the simplest microbes to the most complex plants and animals. Life forms live side by side with non-living physical matter. What makes them different?

Living Things

The world we live in is filled with things that we distinguish from other things. Perhaps the most obvious difference is between things that are alive (like plants and animals) and those that are not (such as rocks and water). Explaining this difference is of paramount importance if we are to account for what, and who, we are, since biological existence enables all other realms of our existence.

Aristotle had a solution to the difference between living and non-living things: living things have souls. He did not mean soul in the way it is typically used today—as an immortal spiritual essence tied to the body during life. Aristotle was talking about something in the body that gives it life and that allows the body to move itself around in the world to sustain itself. For example, plants, he said, have a *vegetative soul* that allows them to grow and reproduce. Animals also have a vegetative soul for growth and reproduction, but in addition, have an *animative soul* that makes possible locomotion in space (behavior) and perception (sentience). Humans not only possess vegetative and animative souls, but in addition, an *intellectual soul* that speaks, reasons, and deliberates.

Galen, the ancient Roman philosopher and physician, expanded on Aristotle. He used the word *pneuma* (from the Greek, meaning breath) to account for Aristotle's soul idea. According to Galen, pneuma, breathed into the body by the lungs, is transported throughout the body, giving it life.

During Galen's time, Christianity was growing in popularity in the Roman Empire, and his Aristotelian view of life, with a soul or spirit in the body, was transformed into the Christian notion of an eternal soul that was separate from the body. For the next thousand years, the Church dominated life in the empire, which had spread across what would be come to be called Europe. By the fifteenth and sixteenth centuries, though, the authority of the Church was beginning to be challenged by the scientific revolution, which offered a belief system based on evidence rather than faith.

The trend toward evidence-based approaches continued in the seventeenth and eighteenth centuries, during the period known as the Enlightenment. But it progressed slowly, in part due to the uncomfortable relation between faith and science. For example, René Descartes, who played a key role in ushering in the Enlightenment, had a strong religious bent, and struggled to reconcile his commitment to science with his faith. His solution was to propose that humans exist in two ways, each dependent on a different substance.

The body, Descartes said, is made of material substance, and its mortal life is a purely mechanistic product, albeit an especially complex product, of the laws of physics. The eternal soul, by contrast, is in his reckoning an immaterial substance that includes a spiritual connection to God, and a consciousness that interacts with the material body. While humans have both a material, mechanical, mortal body and an immaterial, immortal soul, animals have only bodies.

Because he called on two substances, mind and body, to explain human existence, Descartes was a dualist. In particular, he was an interactive dualist because he believed that the soul, via its consciousness, interacts with the body. The locus of interaction, he said, was the pineal gland, which he thought connected the soul to the brain, and hence to the rest of the body.

John Locke was like Descartes—on the fence, tilting toward empiricism and science while remaining religious. Isaac Newton, who revolutionized physical science with his mechanical explanation of the universe, was also a devout Christian who believed that God was the creator of the mechanisms that he and other scientists were discovering.

Later in the Enlightenment, science became more secular, and mechanistic explanations that left out God altogether proliferated. But opposition to using purely mechanistic principles to explain life also flourished. The pushback came not just from the Church, but also from within science itself.

How Vital Spirits Haunted Biology

In the eighteenth century, a movement called *vitalism* gathered steam in the relatively young field of biology. Vitalists believed that some special non-physical quality, some vital spirit, makes living matter different from non-living matter. Although vitalists assumed that a non-physical essence, a vital spirit, underlies life, they were not necessarily spiritualist in a religious sense.

Many biologists in the eighteenth and nineteenth centuries were actually vitalists. Some, such as Louis Pasteur and Johannes Müller, were quite renowned. Pasteur's discoveries helped lead to the creation of vaccinations and the "pasteurization" of milk to thwart disease. He also worked on fermentation, and adopted the vitalist belief that this process could be achieved only using organic (life-based) matter because it required a non-physical force that exists only in living things. He was proved wrong when other scientists showed that fermentation could be achieved with inorganic chemicals.

Another important vitalist was Xavier Bichat, an eighteenth-century French pioneer in the emerging field of experimental physiology. He used the methods of this field to analyze the structure and function of body tissues, and his research marked an important transition toward modern biology. His proposal that each of the twenty-one body tissues he identified is the locus of a unique life force marked him as a vitalist.

The distinction between organic and inorganic chemistry was originally based on the assumption that living things, organisms, were made by organic chemical reactions based on some secret, life-giving force. It wasn't until the discovery that urea, a key ingredient in urine, could be generated from inorganic chemicals that organic chemistry was redefined as being about carbon compounds instead of living matter. This shift was a major blow to vitalism.

The Internal Milieu

The nineteenth-century French physiologist Claude Bernard, who was greatly influenced by Bichat's work on tissues, is recognized as one of the most important experimental physiologists of all time. From his research, Bernard concluded that just as the external environment surrounds our body, the cells that make up the body's tissues are bathed and nourished within a fluid-filled inner environment, which he called *le milieu intérieur.*

For Bernard, "Constancy and stability of the internal environment is the condition that life should be free and independent"—that is, a steady state is required for human wellness. His research showed that the body's internal fluids have a more or less constant ionic (salt solution) composition, and when the external environment changes, compensatory changes take place in the internal milieu to keep the body in balance with respect to its surroundings. To explain the origin of this physiological situation, Bernard proposed that the ability of ancestral organisms to leave the oceans required that they develop a way to "carry the ocean with them," so that their cells came to be bathed in ionic fluids that resemble the very seas from which they evolved.

Darwin's *Origin of the Species* did not impress Bernard. A consummate experimentalist, he thought that Darwin's book was too speculative, too unsupported by data. In stark contrast with the Darwinian view that species were distinct from one another, Bernard assumed that all species had the same basic properties that differed only in degree, and that one could study any animal to understand any problem in biology. He dismissed clinical studies of humans as being too crude, proposing that everything could be learned from more precise studies of other species. Human epidemiology, he argued, was not scientific. Because of Bernard's stature, this negative attitude toward clinical research in humans followed the field of physiology into the twentieth century and hampered the development of treatments for diseases such as polio.

Bernard's anti-Darwinian perspective has been said to have been consistent with the views of creationists of his time, who considered differences between animals to be evidence for small variations on a common blueprint, with the main difference between humans and other animals being a soul. Bernard has also been accused of tilting toward vitalism. But in fact, he was neither creationist nor vitalist. Instead, he challenged the role of both religion and vitalism in biology, writing:

> There are some who in the name of vitalism formulate the most erroneous judgements . . . They consider life a mysterious and supernatural force which acts arbitrarily and releases itself

> from any determinism. . . . vitalistic ideas . . . are nothing more than faith in the supernatural.

Origin of Life

Outside the Darwin-resistant fringes of experimental physiology, evolution by natural selection became the scientific explanation of the relationship between species. And Darwin's proposal that all life spang from some physical event in a primordial pond billions of years ago also provided a scientific hypothesis for how life might have emerged from non-living matter.

With the discovery of the replication capacities of DNA and RNA in the twentieth century, origin-of-life theories became more sophisticated. Two competing accounts emerged. One was that self-replicating molecules preceded life and came to be compartmentalized within a non-biological casing that allowed the proteins they made to be confined within the casing and used to sustain metabolism. With time, these so-called protocells evolved their own biological enclosure, a membrane, and true cells were born. To stay alive, these early cells had to balance their inner milieu relative to the outside environment in the process of making energy through metabolism. And they had to use the energy they made to sustain life and to enable reproduction by DNA replication. The other leading theory has the same major chemical players, but with metabolism preceding replication in the historical sequence.

Regardless of which model is correct, once cells existed, the driving force that allowed each individual cell to persist (to live) was metabolism, while replication was required for the species to persist beyond the lifespan of the individual cell. Metabolism and replication are defining features of life. Everything else about living things rests on the foundation of these two processes.

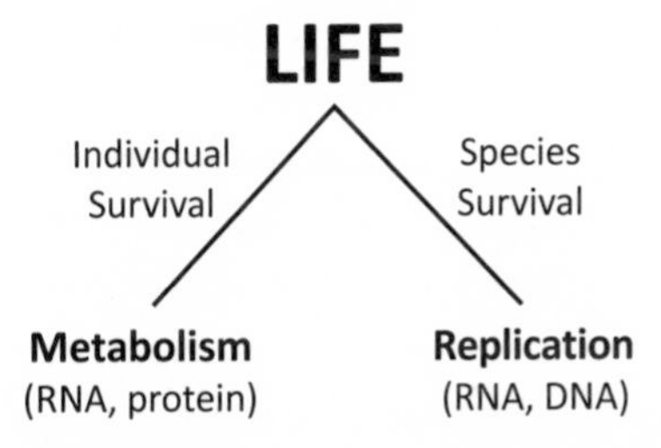

Figure 6.1. Metabolism and replication are fundamental requirements of life. Based on Paul Cisek, "Resynthesizing Behavior through Phylogenetic Refinement," *Attention, Perception, & Psychophysics* 81 (2019): 2265–2287, figure 8.

Homeostasis, Metabolism, and Life Itself

Bernard's ideas about the stability of the internal environment were reconceived in the late 1920s by the American physiologist Walter Cannon. While Bernard had shown that the body maintains a constant inner milieu, Cannon marshaled evidence to explain how this stability between the inside and outside is achieved.

Cannon focused on the role of adrenaline, a hormone released from the adrenal medulla in so-called emergency situations. By binding to receptors on various inner organs, adrenaline increased those organs' activity to meet the demands of the emergency, in part by releasing energy supplies from storage in the liver or in fat. Once the emergency was over, additional physiological changes were required to restore balance. He called this process *homeostasis.*

Although Cannon, and many who followed in his footsteps, focused on how the body responds to harmful, extreme conditions, the implications of the concept of homeostasis are much broader. They are about how the body stays alive by continuously adjusting, moment-to-moment, under routine conditions, as well as in emergencies.

Allostasis is an important contemporary notion related to homeostasis. It refers to the body's ability to predict future needs and make anticipatory adjustments. Homeostasis restores balance, whereas allostasis seeks to maintain optimal functioning through anticipation.

A key role of homeostasis is ensuring a chemical milieu suitable for metabolism. In eukaryotic cells, metabolism is the job of tiny energy machines called mitochondria. Each cell in the body makes its own energy by using oxygen to break down glucose. To achieve this outcome, the fluids surrounding the cells must have a suitable concentration supply of glucose and oxygen, as well as ions (calcium, potassium, sodium), minerals (iron, magnesium), and a proper pH (acid-base balance). These chemicals are delivered throughout the body via the bloodstream. Blood flow requires that the heart be beating. And because the chemical reaction that breaks glucose down requires a narrow temperature range, this too has to be precisely regulated. Each of these steps is controlled by feedback processes that stabilize the internal milieu, making it possible for metabolism to efficiently produce energy. The health and vitality of the organism depends on metabolism and homeostasis, which together are referred to as *metabolic homeostasis.* The renowned scientist Bruce McEwen, writing with colleagues about the role of metabolism in life, put it this way:

> Life emerges when biological structures are animated by energy. Energy is defined as a fundamental entity of nature that is transferred between parts of a system in the production of physical change within the system, and usually regarded as the capacity for doing work. . . . Without energy, there is no life—molecules alone do not interact in meaningful ways, and complex structures do not assemble nor replicate. . . . it is a required quality of living organisms to experience a constant flow of energy through and between their different parts. Without energy . . . the body dies.

Echoes of Vitalism Today

The life-giving and life-sustaining energy of metabolic homeostasis is at the core of the biological realm of existence. Metabolism is a far cry from the mysterious, nonmaterial forces proposed by vitalists. Yet despite the achievements of science, vitalism continues.

Today, vitalism is a tacit manifestation of spiritualism, with various terms used to name non-material essences that are assumed to be part of life. Some terms come right out the vitalist tradition (*vis essentialis, vis vitalis, élan vital,* vital impetus, life force, life spirit), and others are associated with Greek philosophy, Christianity (soul, spirit), or Eastern spiritualism (chi, brio, ki, prana, shakti, purusha, jōriki, astral body).

Belief in a soul or afterlife, or in spiritualism in general, will probably always exist among humans. That is not my concern. What does trouble me is that such ideas are sometimes treated as legitimate scientific explanations. Vitalism, as Bernard said, has no place in science.

Some scientists are deeply religious, or spiritual in other ways. But as Cicero and Locke, and many others since, have said, we assume different roles (personas) in different situations. So long as scientists are scientists when they do science, they can lead their non-scientific lives as they choose, and no harm is done. But when they use their scientific credentials to give credence to non-scientific, and even anti-scientific, points of view, harm most definitely can result.

Things can also get tricky when scientists interact scientifically with philosophers who hold positions such as mind-body dualism or panpsychism (the belief that consciousness permeates all of nature, living and inert). These are perfectly legitimate philosophical areas of inquiry, but that doesn't make them legitimate scientific topics.

Dualism and panpsychism have both been linked with vitalism, and some scientists have developed scientific theories compatible with these philosophical views. For example, a position called *integrated information theory* has been said to be a panpsychist model of consciousness. It proposes that an entity called *phi* reflects the synergy and information complexity of a physical system. According to this theory, any physical system that stores information has a phi value, and therefore some degree of consciousness. Because phi, and consciousness, are spread throughout the universe, this theory holds that stardust, rocks, and tables—and all manner of other objects—are composed of bits of consciousness.

Considerable pushback has been directed at integrated information theory, with opposing scientists arguing that the proponents are flirting with ideas that are, at best, on the fringe of scientific practice, in part because they are shrouded in a veil of complex math, but also because they seem untestable and not subject to refutation. Some have said that integrated information theory requires an unscientific leap of faith, while others have called it pseudoscience.

Like most biological scientists, I believe that life emerged some four billion years ago as a result of a perfect chemical storm. All living things, what are called organisms, exist in this biological realm of existence. The secret of organismic life is not some mysterious vital or spiritual force. It is instead the mundane processes of metabolism and replication that have made possible the life of every organism that has ever existed.

7

Bodies

Plato is said to have encouraged philosophers to look forward to death and the release from bodily passions and other limitations. But for better or worse (depending on your perspective), modern science does not support the idea of a Platonic pure state of existence that persists beyond the last beat of one's heart. All aspects of a living thing depend on and die with its body. Sorry Plato, when your troubling biological realm ceased, your nervous system shut down, which meant that your cognitive and conscious realms, and hence your capacity for reason, were eliminated.

Organisms

In *The Wonders of Life,* published in 1904, the prominent evolutionary biologist Ernst Haeckel provided this definition of an *organism:*

> In the sense in which science usually employs the word "organism," and in which we employ it here, it is equivalent to "living thing" or "living body." The opposite to it, in the broad sense, is the anorganic or inorganic body. Hence the word "organism" belongs to physiology, and connotes essentially the visible life-activity of the body, its metabolism, nutrition, and reproduction.

To this definition, we can add that organisms change over time through physiological growth and repair processes controlled by their genes. Genetic activity, in turn, is modulated by the environment in which the growth or repair is occurring. Two genetic properties are especially important: uniqueness (the genome of each individual differs from that of all other individuals) and homogeneity (all cells of an individual organism share a common genome).

The way the particular combination of genes is expressed in the building of an individual's body makes it a unique specimen within its species, in terms of both its external appearance and its internal cells and their physiological functions. Because of genetic homogeneity, all of the cells within an organism are physiologically compatible. As a result, organisms are able to operate internally with a high degree of cooperation and a low degree of physiological conflict as they work to make energy through metabolism and use energy for homeostasis and reproduction. Illness affecting cells in one part of the body, if severe, can disrupt this cooperation and induce sufficient physiological conflict so as to impair the organism's unity or even threaten its physiological viability.

Three key features of organisms were identified by the Chilean biologists Humberto Maturana and Francisco Varela: self-production (generation of one's own parts); self-organization (correct assembly of one's own parts); and self-maintenance (taking care of one's own nutrition and repair). Like organisms, machines are complex entities that function as a unit, and both are open systems with self-regulatory processes that use energy to do work. But machines are designed, assembled, and repaired by external agents, usually humans. And because they are physical, but not biological, machines lack metabolism, growth, self-production, self-replication, and self-repair.

A machine with a program exists as a machine with a program; nothing more. We humans exist and function the way we do because we are the result of a 3.7-billion-year evolutionary history of gradually accumulated and intricately entwined changes in body design, with each new step building on previous ones.

What about machines under the control of artificial intelligence? These have some features that are closer to those of a complex animal, such as the ability to gather information and use it to learn rules and follow them, and to engage in relatively simple forms of pattern recognition, planning, and decision-making. Some say that such machines have cognition, and might have consciousness. It has even been proposed that so-called cyborgs are (or will be) *post-human persons* with ethical and moral responsibilities and rights. I believe, however, that these AI-driven machines, because they have not evolved a biological and neurobiological way of life, and have not gone through a maturation experience that instantiates the biological and neurobiological processes that constitute an individual of that kind, are, as their moniker *artificial* implies, merely mimics of actual cognition (for example, chatGPT is a pretty good mimic of human cognition), rather than primitive exemplars of it. More on AI at the end of this book.

Organisms as Biological Individuals

The organisms we share the planet with look the way they do because they inherited the body plan that their species evolved through natural selection. In other words, natural selection alters the bodies of organisms.

As articulated by Richard Lewontin, natural selection works this way: Variation exists between individual organisms in a population of similar organisms, and when individuals mate and reproduce, their offspring inherit traits that make them more or less fit in their environment. If the environment (which includes not just the physical environment, but also other organisms) changes, individuals with traits useful under the new conditions become more plentiful. Over time, if the species continues to change in response to environmental pressures, it will diverge further from its ancestral group, and at some point, via the accumulation of changes, a new kind of organism, a new species, will come to exist.

Leo Buss expanded on the traditional Darwinian approach, putting natural selection into the context of a hierarchy of changes by which new kinds of organisms arise. New organisms, he pointed out, not only possess novel features, but also retain features of the group they diverged from. His crucial insight was that new (or newly changed) features often become the primary target of natural selection, with selection based on old features being less common, and in fact, suppressed. For example, the basic life-sustaining physiological functions inside the bodies of mammals have been repeatedly tested for their survival value by natural selection, and therefore tend to change relatively little in the evolution of new species. More often, the changes involve processes that control the way the organism's particular kind of body interacts with its environment in satisfying its life-sustaining inner needs.

Buss did not pull this out of a hat. Decades before him A. B. Novikoff had written:

> The concept of integrative levels of organization is a general description of the evolution of matter through successive and higher-orders of complexity and integration. . . . [N]ew levels of complexity are superimposed on the individual units into a single system. What were wholes on one level become parts on a higher one. Each level of organization possesses unique properties of structure and behavior which, though dependent on the properties of the constituent elements, appear only when these elements are combined in the new system. . . . "Mesoforms" (i.e. intermediate forms) are found at the transition point of one level to the next.

Building on such ideas, John Maynard Smith and Eörs Szathmáry sought a mechanistic, and specifically a genetic, explanation of how the transitions from lower to higher levels took place over the course of evolution. They proposed that the two key genetic features discussed earlier (homogeneity and uniqueness) could be used to assess changes in variation and selection at major transition points.

Paraphrasing the evolutionary biologist Richard Michod, some of the major transitions, starting with the existence of self-replicating genes, were: genes → gene networks → protocells → cells with free-floating DNA (prokaryotes, such as bacteria) → cells with sequestered DNA that sexually reproduced (eukaryotes such as protozoa, algae, and amoeba) → multicellular eukaryotes (plants, fungi, animals).

Evolutionary theorists often describe the relationship between ancestral groups and their descendants in term of *nested hierarchies.* For example, eukaryotes evolved from prokaryotes, with new features supplementing the old, and the cells of multicellular organisms are all eukaryotic.

It's worth emphasizing that each of the three multicellular groups evolved from a different unicellular eukaryotic ancestor. For example, protozoa were the unicellular ancestors of animals. This is a crucial point for understanding how present-day animals, which are our focus in this book, came to exist, and how protozoa connect animals, including humans, to the earliest life forms that emerged billions of years ago.

Biological Individuals Reconsidered

The views considered so far assume that the organism is the target of natural selection and, hence, the basic unit of biological individuality. Although different criteria may be needed to define organisms that emerged at different transition points, the organism, whatever it is, according to the traditional view, is the target of natural selection.

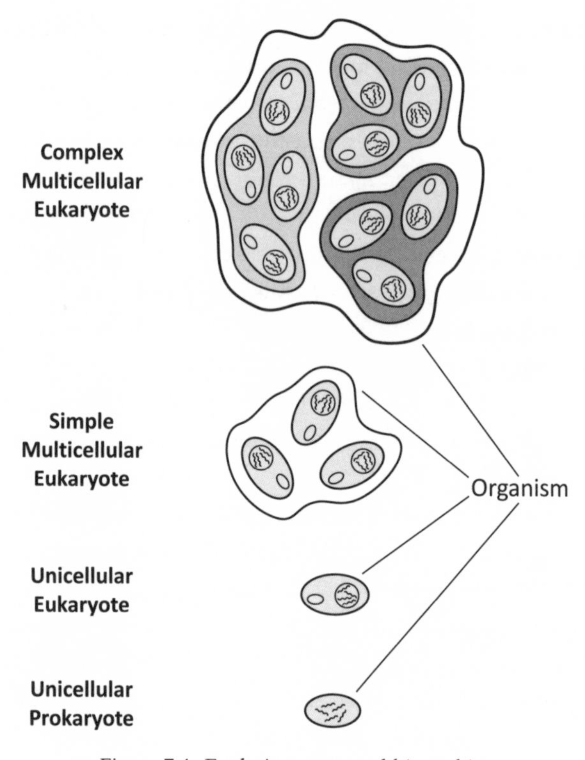

Figure 7.1. Evolutionary nested hierarchies

But this focus on the organism has been questioned. For example, the philosopher of biology David Hull has challenged the traditional equivalence of the terms *organism* and *biological individual*. He notes that although organisms are integrated functional units, they are just one component of a hierarchy of biological individuals that includes species, populations, individual organisms, cells, molecules, genes, and chromosomes. The individual organism is merely a middle level in Hull's hierarchy of biological individuality.

An important part of Hull's argument is the fact that within a given organism there are sometimes other biological individuals. A baby in the womb of its mother is an example. The microbiota (bacteria) that live in our digestive system and that support our health and well-being are another instance (we actually have many more bacterial cells in our body than human cells). These visitors, which share physiological mechanisms such as metabolism with their host, are called *symbionts.* The resulting combined biological individual is known as a *holobiont* and its combined genome is called a *hologenome.*

Not all outside visitors are welcome additions. The bacterium *E. coli* and the single-cell protozoan parasite *Giardia* can both wreak havoc on our digestive systems. Although viruses are not usually considered living things since they do not replicate on their own, they take up residence in biological individuals and replicate using the host's DNA, affecting physiological well-being. Also, when the cells of an organism become cancerous, they cease to work for the greater good of the body and become in effect unwitting defectors from the integrated physiological organism. It has even been proposed that cancer occurs when the cells of a multicellular organism revert to a single-cell mode of independent existence, which is supported by studies showing enhanced expression in human cancer patients of genes that we humans share with single-cell eukaryotes.

Darwinian Individuals

Hull's challenge to the traditional view of organisms as the *sine qua non* of biological individuality has led to a debate about how to best distinguish and define different kinds of biological individuals. The philosopher Peter Godfrey-Smith, a major player in this debate, focuses on what he calls *Darwinian populations.* These are groups with variable heritable traits, some of which provide reproductive advantages to the population members that possess them. Members with these traits are known as *Darwinian individuals.* But for Godfrey-Smith, Darwinian populations and Darwinian individuals are not necessarily organisms, since he also allows fragments of organisms (genes and chromosomes, for example) to qualify as Darwinian individuals.

This model differs from Richard Dawkins's well-known theory of the selfish gene, which is also based in part on Darwinian principles. For Dawkins, genes are replicators that are housed in organisms. The survival of replicators is, in Dawkins view, what matters in natural selection, because they ensure high-fidelity gene transmission across generations. By contrast, Godfrey-Smith shifts the emphasis from replicators to the process of replication (reproduction). For him, reproductive success (fitness) is the target of natural selection. Organisms that do not contribute to reproductive fitness, he says, are not Darwinian individuals, even if they live among Darwinian individuals within a Darwinian population. This seems like a harsh assessment of people who choose not to, or are unable to, have children, but contribute to the viability of the species in other ways.

Physiological Individuals: The Case for Integrated Metabolic Activity

The philosopher of science Thomas Pradeu is a leading proponent of the idea that *physiological individuality* determines biological individuality. This position, by contrast with the notion of Darwinian individuals, places the organism front and center. Pradeu emphasizes the importance of integrated metabolic activity as a major factor underlying the unity of function of biological individuals. His model specifically proposes that constant systemic immune reactions control metabolic activities in the body moment-to-moment, and these, he says, ensure the unity of the complex heterogeneous physiological processes that sustain the life and well-being of the individual.

Most striking is Pradeu's claim that the fundamental boundary of the individual organism is not an anatomical structure, like skin or other kinds of covering, but a chemical boundary maintained by the immune system. In his words:

> The immune system plays a key role in monitoring every part of the organism and maintaining the cohesion between the components of that organism, making each individual unique and constantly re-establishing the boundaries between the organism and its environment.

Pradeu's view goes against the dominant position in immunology, the assumption that the immune system, by recognizing and rejecting foreign antigens, distinguishes self from non-self. He says this can't be correct because the immune system tolerates symbionts such as the microbiota in our gut.

Although Pradeu agrees with Hull and Godfrey-Smith that the expression *biological individual* refers to more than just the organism, he argues that among biological individuals the organism is the most integrated form of biological individuation. He says the presence of a fetus or microbiota in an organism does not invalidate the idea that the human organism is the core of *human biological individuality.* Pradeu also points out that to know what kind of evolutionary (Darwinian) individual a biological entity is often requires knowing what kind of organism it is.

An influential article by Robert Wilson and Matthew Barker points out that, regardless of the eventual status of Pradeu's provocative immune hypothesis of individuality, his focus on the organism as the unit of biological individuality is a welcome redress to the skewed focus of the recent literature on Darwinian individuals. I agree completely; this book is unapologetically about organisms.

Organisms as Chemistry Sets

The biological life of an organism, whether unicellular or multicellular, can be characterized as a set of continuous chemical interactions. For example, the life of a unicellular organism depends on chemical reactions within the cell (such as metabolism), but also on interactions between the chemistry of the cell interior with the chemistry of its local environment (such as homeostasis). Notably, the external environment typically includes other cells that are also interacting homeostatically with that same environment and with each other.

The transition from unicellular to multicellular eukaryotic life required a solution for the problem of how the cells could interact to maintain the unity of larger, more complex organisms. Some basic principles of cell-to-cell interaction were worked out by cells that adhered to one another as colonies. These are not true multicellular organisms, however, as they are simply aggregations of convenience. For example, by colonizing, eukaryotic prey could achieve safety in numbers from eukaryotic predators. Another survival advantage of colonization results from a division of labor. A single cell living solo has to control movement, digestion, reproduction, and so on. In a colony, different cells take on different tasks. This division of labor is achieved by the various cells releasing chemicals that inhibit some of the normal functions of other cells by altering gene activity in the group. The resulting specializations, though, are opportunistic and transactional. If a cell defects from the colony, its genes resume their normal functions and the cell survives solo.

Prokaryotes (for example, bacteria) and eukaryotes (for example, protozoa and algae) form colonies, but only eukaryotes found a way to exist as true multicellular organisms. In fact, true multicellular organisms evolved via transformations of unicellular colonies.

A true multicellular starts life as a single cell (fertilized egg), the mother of all other cells that will come to be the organism. The multicellular organism begins to be assembled into an integrated body during embryonic development by a genetically encoded body plan that results in part from the mixing of genes from two individuals via sexual reproduction, a capacity that first appeared in eukaryotes. By the sheer math of the mixing, each individual has a different body compared to all other species members, and all species bodies differ from the bodies of all other species. A multicellular organism is a functional unit from the start to the end of life, such that the various cells can exist only as part of the larger unit—a skin cell cannot separate from the rest of your body and survive on its own in the world.

The specific life-sustaining chores of the body of a multicellular organism are performed by genetically specialized cells (for instance, in animals, skin, blood, heart, lung, muscle, and immune cells) that make up the tissues and organs of systems. Chemical interactions take place not only between cells near each other in the tissues of a system, but also between cells in different systems. For example, oxygen extracted from the air by your lungs is carried in the bloodstream to various tissues and to cells throughout the body for metabolism. The bloodstream also transports hormones produced in the endocrine system throughout the body, allowing these chemicals to play an important role in coordinating the activities of the whole organism.

This description is especially apt for animals like us, but also applies, in a general way, to other multicellular organisms. For example, minerals and water extracted by the roots of a plant flow through a vascular system to the leaves to support metabolism.

Organisms Change but Remain the Same

The organism is always in flux, constantly adjusting metabolically, homeostatically, and / or allostatically every moment of every day. In early life, these responses are preprogrammed, automatic, and rigid. As development proceeds, the body has to adapt to emerging capacities and conditions that result from maturation and growth, and from changing circumstances. With time, capacities to learn and adapt proliferate, allowing the acquisition of information about the world, and one's relation to it. Later in life, the organism has to cope with diminished capacities (in humans, for example, hearing loss, cataracts and other visual problems, heart disease, memory impairments, cancer, and so on).

That organisms constantly change is why philosophers and scientists have struggled for so long with questions about individuality and identity—how do we remain the same as we change? Derek Skillings has referred to individuality and identity as two sides of the same coin. Individuality, which has been the focus of much of this chapter, is about boundaries—what makes biological individuals different from the start? Identity, on the other hand, is about what makes a given biological individual the same over time.

Genes, immune responses, and other biological properties both help make each individual biologically different, and contribute to that individual's biological sameness or identity over time. Yet neither individuality nor identity are solely biological conditions for humans. Neurobiological, cognitive, and conscious factors also make important contributions to what, and who, one is. Cultural factors contribute, too, as cognitive and conscious constructions.

8

The Duality of Biological Existence

My idea about us existing in different ways was not the result of a de novo, out-of-the-blue insight. Scholars have long understood that animals exist in different ways simultaneously. For example, as I explained earlier, Aristotle said that animals exist both vegetatively and amatively—with vegetative functions being about the inner body and animative function being about interactions with the outer world.

Building on this distinction, Bichat, the eighteenth-century tissue physiologist (and vitalist), proposed a similar duality of animal existence. According to his view, animals, through their so-called vegetative inner tissues, live a *vie de nutrition* (life of nutrition) and through their sensory and motor tissues, a *vie de relation* (a life of behavioral interactions with the world). Bernard expanded on Bichat, proposing internal and external milieus, and Cannon added homeostasis as the way the inner and outer worlds are kept in balance. By contrast with Descartes's philosophical separation of a physical (biological) body from a non-physical mind, the different kinds of animal existence just mentioned were simply biological.

There's no shortage of material on this topic. For example, contemporary biology and neuroscience textbooks devote much space to the physiology of the inner tissues and to the sensory and motor mechanisms that animals use to behave in the world. But there's a particular perspective on this duality of animal life that had a major impact on how this book unfolded.

Romer's Body Duality

Early on in writing this book, I stumbled across an article by the paleontologist Alfred Sherwood Romer that was initially published in the 1950s, then re-published in the 1970s in a book on evolutionary biology. It escaped my notice, and, as far as I can tell, that of most neuroscientists, for the next forty-eight years.

It's not that Romer himself was overlooked. He was widely recognized for his work on the anatomical changes required for evolutionary transition from fish to amphibians, and has been described as one of the important evolutionary anatomists of the twentieth century. His 1955 textbook, *The Vertebrate Body,* is considered the bible of comparative anatomy. Yet the particular paper in question seems to have seldom been cited in the mainstream literature on brain anatomy, and citations that did occur were mostly in technical papers on very specific topics. For me, though, it ended up becoming a crucial part of my conception of this book.

I have to admit, it took several readings. At first it seemed like a straightforward reworking of the ideas of Aristotle, Bichat, Bernard, and Cannon. But the more I delved into to Romer, the more interesting I found what he had to say.

Romer focused on two classes of muscle tissues known to be possessed by vertebrates. Striated muscles make up much of the body's flesh and are attached to the skeleton. These respond rapidly in controlling body movements in the external world, and they underlie behavior. Smooth muscles, by contrast, are traditionally associated with inner organs and glands that take care of digestive, respiratory, reproductive, and other physiological functions. These contract more slowly than striated muscles.

The difference between the two is related to differences in the kinds of proteins that control the contractions. The mix of these proteins in striated muscles underlies their contractile efficiency, and the way the proteins are arranged gives the appearance of visible bands, striations, that are the basis for their name.

But Romer noted that the association of striated muscles with skeletal movement and smooth muscles with the inner tissues was an oversimplification. In particular, in some vertebrates, some smooth muscles have striations. While the existence of these hybrid (striated / smooth) muscles was known for some time, it was treated as a minor anomaly. Romer, however, saw a significant evolutionary process at work.

Like Bernard, Romer built on the importance of the vertebrate transition from an aquatic to a terrestrial lifestyle. He proposed that in order to meet the metabolic challenges imposed by the new environment, early amphibians needed more efficient muscles in the heart, lungs, and digestive track. This led to the addition of rapidly contracting proteins in some smooth muscles, which gave those muscles striations similar to skeletal muscles.

Romer concluded that since the presence of striations is not limited to muscles that move the skeleton in space, the way to think about the relation of muscles to body function is not in terms of the distinction between striated and smooth categories. Instead, he said, the key distinction is between somatic (behavioral) and visceral (so-called vegetative) body systems, regardless of the muscles those systems use. The implications of Romer's conclusion are perhaps best left to his own words:

> In many regards the vertebrate organism, whether fish or mammal, is a well-knit unit structure. But in other respects there seems to be a somewhat imperfect welding, functionally and structurally, of two somewhat distinct beings: (1) an external, "somatic," animal, including most of the flesh and bone

> of our body, . . . and (2) an internal, "visceral," animal, basically consisting of the digestive tract and its appendages, which, to a considerable degree, conducts its own affairs, and over which the somatic animal exerts but incomplete control.

On Romer's Terms

I think Romer used *visceral* instead of Aristotle's more widely used *vegetative* term because, from a tissue perspective, it better differentiated the actual kinds of muscles at work in the two kinds of activities. To *eviscerate* means to remove the innards, or *offal* as they are called in culinary circles. Scientifically, the viscera include three kinds of tissues. The first are the major organs (heart, lungs, gut, liver, gall bladder, kidneys, and brain). The secretory tissues that release hormones and fluids are the second (adrenal medulla, adrenal cortex, pancreas, gonads, thyroid, parathyroid, pineal, salivary, lacrimal, and sweat glands). The third are blood vessels (arteries and veins). Most visceral tissues are located in the cavities of the abdomen, chest, and pelvis, though a few are in the neck and head regions; and the blood vessels run through all body tissues, including bones. Visceral tissues keep us alive. They take care of the life-sustaining homeostatic processes that allow metabolism to function in all the cells of the body, moment to moment.

Besides these scientific reasons, there is a social reason to avoid the term vegetative. It has come to name a neurological condition, *persistent vegetative state,* that occurs when the lower body's vegetative-tissue functions are still capable of sustaining life, but because of brain damage, the patient is not awake and responsive. This term is insulting, because it implies the patients are vegetable-like. Use of *persistent visceral state* would carry no such stigma.

I have explored Romer's ideas in such depth here because I ended up gleaning something very important from him: not only are visceral and somatic functions two subcategories of the biological realm, they are also the foundation on which all of the other realms were built evolutionarily. This will become apparent in Part III, when we explore how the nervous system, and hence the neurobiological realm, evolved from non-neural tissues.

PART III

THE NEUROBIOLOGICAL REALM

9

It Took Nerve

In a short span, overnight in evolutionary time, life went from being solely unicellular and microscopic to also being multicellular and macroscopic. While the emergence of each of the three macroscopic kingdoms of life—plants, fungi, and animals—was clearly a significant evolutionary event, the transition to animals was especially important.

An Extraordinary Transition

Animals were the last of the multicellular kingdoms to appear, arriving around 800 million years ago. Armed with a game-changing feature, they developed unparalleled abilities to assess the world around them and respond adeptly to challenging situations in novel niches. That feature was a nervous system.

In many ways, the nervous system is just another body system: the heart pumps blood, the lungs extract oxygen from the atmosphere, and the nervous system controls behavioral interactions with the environment. This neural control over animal behavior is so commonplace that we sometimes take it for granted. But consider how remarkable it is that you can rapidly withdraw your foot from a sharp rock to minimize the degree to which it penetrates. During that fraction of a second, the stimulus must go from sensors in the skin on the bottom of your foot, up your leg to your spinal cord, and then back down your leg to the muscles that, by lifting the leg, pull your foot away. Or how about a frog's nervous system, which can compute the trajectory of a fly that is moving through its field of vision, project and retract its tongue, and begin digesting that fly by the time a human observer knows what is happening?

Having a nervous system greatly enhanced animals' ability to sense and respond to not just the outer world, but also their inner tissues. And it offered animals a novel, in fact, revolutionary, way of surviving and thriving in the world as they went about satisfying the metabolic requirements of life. Over time, the synergy between sensing and reacting evolved into learning; learning evolved into associative learning; associative learning into thinking; thinking into knowing; and knowing into experiencing.

Without a nervous system, life would be merely biological, because there would no neurobiological realm of existence, and, as a result, neither a cognitive nor a conscious realm.

The Animal Way of Life

Animals, as a rule, survive by searching for and consuming nutrition, and by avoiding being a source of nutrition for other creatures. While the protozoan ancestors of animals were acutely sensitive to nutrients and danger, and they were behaviorally active in approaching useful substances and withdrawing from harmful ones, their stimulus detection and response capacities were limited to what their single-cell bodies could do.

Coordinating the movements of a multicellular organism in space is a much more complex process, because the cells that detect external stimuli are not necessarily adjacent to the cells that control the muscles that move the body. The evolutionarily old-fashioned means of internal communication used by unicellular microbes in colonies—the diffusion of chemicals between cells—is too slow for communication within multicellular organisms that make their living by responding with speed and precision.

Animals became speedy responders by using two innovations. One involved sprouting an extension, known as an axon, that could put cells in touch over long distances. The second innovation was to use electrical signals to rapidly zip through the axon at speeds that can exceed two hundred miles an hour. Once the action potential reaches the end of the axon, it causes the release of a chemical neurotransmitter, which crosses the small space (*synapse*) to reach receptors on the surface of the receiving cell. This process generates small electrical impulses in the receiving cell, and if enough electricity is created in a short enough time, an action potential is produced in it, so that it can continue the chain reaction. Cells that have these features are called neurons.

Action potentials did not have to be invented. They existed for other purposes in unicellular eukaryotes, and just had to be put to use in animal nervous systems as a means of fast transmission within cells. But rapid transmission within the nervous system is only useful if the body itself can respond rapidly.

The protozoan ancestors of animals used cilia to move around, rapidly but randomly, and to turn toward nutrients or away from toxins when those were detected (bacteria use flagella to behave similarly). Primitive, stimulus-guided behaviors like these are called taxic responses. But for sheer physical reasons, cilia are not a practical way to move a large multicellular body. The contractile properties of protozoan cilia were, however, transformed into muscles and neurons in tandem in animals. Stimuli detected by sensors could then trigger action potentials in sequential neurons, allowing the animal to rapidly contract striated muscles when fleeing from predators or when hunting and capturing prey. Animals also developed visceral (smooth) muscles to digest captured food to make energy, manage wastes, and in general, maintain metabolic homeostasis / allostasis.

Building Blocks of Nervous Systems

The first animals, sponges, arrived about 800 million years ago. As animals go, sponges are very primitive because they have limited kinds of body tissues. Although present-day sponges do not possess a nervous system, they have genetic and molecular building blocks that underlie the nervous systems of animals that evolved from them. And sponges got these from their single-cell protozoan ancestors. Both used these building blocks for non-neural purposes. For example, one important building block is a class of cells known as adhesion molecules. These hold individual protozoa together to form colonies; they also hold together the cells that make up body tissues (including neural tissues) in sponges and other animals.

About 700 million years ago, Cnidaria, the phylum that includes jellyfish and hydra, diverged from sponges. These animals are radial in shape, possessing a single axis, a top and bottom, but having neither a front-back nor left-right distinction. They, unlike sponges, have nervous systems. All of this suggests that the key event that made neurons and synapses possible in Cnidaria was not a sudden arrival of novel genes, but instead the use of existing genes in a new way.

The Cnidarian nervous system consists of a nerve net, a collection of interconnected neurons diffusely distributed throughout their skin. That this relationship between neurons and skin lives on is shown by the fact that the nervous system and skin of animals are both derived from the same embryonic layer (the ectoderm) during early development.

Of the Cnidaria, jellyfish have been studied most extensively. They have receptors for light, touch, and gravity in the outer skin of their umbrella-like dome, and underneath are muscles that propel movement. The nerve net interconnects the sensory receptors to the muscle effectors, and controls not only behavior, but also metabolic homeostasis. Jellyfish use their tentacles to hunt, capture, and

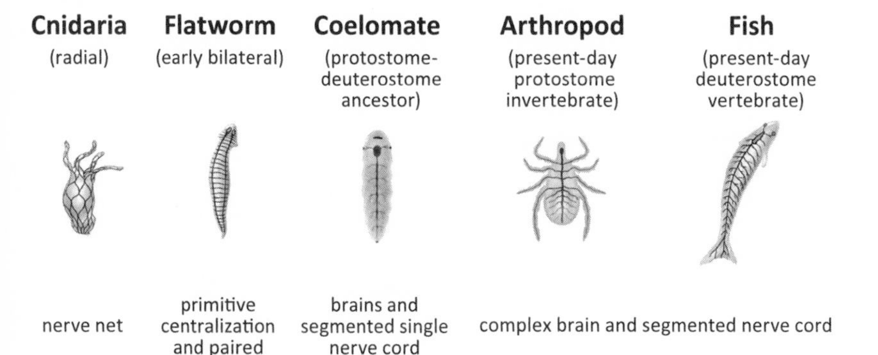

Figure 9.1. The evolution of nervous systems. Reformatted from Joseph LeDoux, *The Deep History of Ourselves* (New York: Viking / Penguin Random House, 2019), figure 30.2.

consume food. The nervous system then facilitates the digestive activities needed to convert that food into energy and to excrete wastes. Sensory-motor integration by the nervous system is also used to escape from predators.

It's important to note that primitive animals like jellyfish don't actually "perceive" the food or the predator as such. They have sensors that automatically respond to chemicals in nutrients by initiating an approach toward the food. And anything that touches their skin triggers an explosive burst of muscle contractions that propel escape. These two kinds of sensory-motor integration, one related to feeding and the other to defense, are in a sense direct extensions of basic processes of approach and withdrawal that are present even in microbial organisms.

Control of these muscle-based approach-and-withdrawal capacities by the nervous systems of jellyfish became the prototype for what more elaborate nervous systems would do in the animals that followed. Specifically, jellyfish have two concentrated collections of neurons within their nerve net—one involving the mouth and tentacles contributes to feeding, and another around the edge of the dome contributes to escape. These collections of peripheral neurons are believed to have been the forerunners of centralized neural circuits, which in turn became the foundation of the central nervous system.

Cnidaria also contributed basic mechanisms of sleep and wakefulness, including the sleep-regulating hormone melatonin, to the animals that evolved from them, including us.

Bilateral Bodies Needed More Than a Nerve Net

The first beneficiaries of the Cnidarian effort toward centralization were flatworms that diverged about 630 million years ago. Their bodies possessed the three axes typical of most animals we live with (top-bottom, left-right, and front-back). This organizational scheme is referred to as *bilateral symmetry*.

The front end of early bilaterals had a head with a face pointing forward—that is, in the direction of its locomotion (see Figure 9.1). It contained a concentration of neurons—a primitive brain. The reason why most animals alive today, including us, have bilateral symmetric bodies with a brain in the head is because we all descended from these early worms.

The two main groups of bilaterals that resulted from the next evolutionary steps are commonly called invertebrates and vertebrates. Vertebrates are bilateral animals with a vertebral column, and include fish, amphibians, reptiles, birds, and mammals. Invertebrates, which lack a vertebral column, include both aquatic animals like worms, mollusks, and crabs, as well as terrestrial animals like worms, insects, and spiders.

Why did bilateral symmetry become so common among animals? The answer is both simple and profound: bilateral symmetry provided an important advantage, and so was evolutionarily preferred. For a radial-bodied jellyfish, rapid and widespread muscle contractions, and the resulting undirected escape response, are enough for survival. But directional control over a bilateral body composed of three axes, and that can turn left when danger appears on the right, or can twist right when food is detected there, is only possible if there is some means to control diverse, spatially separated muscles in response to multiple kinds of stimuli. And that is what centralization provided.

The simple stimulus-response mechanisms of Cnidarians are in effect reflexes—reactions that are determined by the stimulus-response wiring of the nervous system. Reflexes are neural versions of the even simpler feeding-related approach and withdrawal behaviors in microbes, with neural control adding speed and specificity. Reflexes persist in all bilateral animals, and have been tailored to the body plans of each species.

Early bilateral animals evolved more complex kinds of behaviors based on centralized control by a brain. These behaviors are usually called *species-typical fixed action patterns.* Like reflexes, species-typical reactions are automatically elicited by particular stimuli. But they involve much more complex patterns of body movements, sometimes consisting of sequences of responses that are controlled by motor programs in the brain. It is believed that fixed action patterns evolved by modification of reflex mechanisms.

Species-typical reaction patterns can be seen in elaborate brain-mediated behaviors that evolved to support the survival of the species—to help it find and consume food, for example, or maintain a balance of fluids, defend against danger, and reproduce. As with any complex movement, multiple reflexes occur simultaneously to support the execution of species-typical reactions.

Although fixed reactions are traditionally thought of as innate instincts, heated debates have occurred over whether they are truly innate. Today it is accepted that, while there is a strong inborn component to such responses, they can also be influenced by individual experience.

The Cambrian Explosion

The period between about 540 and 480 million years ago is known as the *Cambrian Explosion,* so named because of the rapid and expansive diversification of animals, especially of bilateral animals, during this relatively short time span. All of the major animal phyla we live among today were in existence by the end of the Cambrian, although considerable species diversification within each phylum continued to take place.

Armed with powerful species-typical reactions, Cambrian bilaterals became predators and prey, triggering *evolutionary arms races* in which animals continuously adapted in response to features of other animals in ways that made each more competitive—becoming bigger, stronger, better weaponized or better defended, and able to invade new environmental niches. Predation and defense were likely not what initiated the Cambrian Explosion, but they certainly contributed to its grandeur.

Much of the waking life of animals is spent searching for and consuming food by using, at a minimum, complex combinations of reflexes and fixed reactions. Whether predator or prey, an animal must continuously replenish the nutrient resources needed to maintain metabolic homeostasis. Because food is not always easily available, it must be sought out through foraging.

According to Thomas Hills, bilaterals in the Cambrian employed a very basic and ancient food-foraging strategy. It is called *area-restricted search,* and involves two interrelated components that balance *exploring* a large area against *exploiting* local resources. The exploring component uses fast movements to do a cursory search over an extensive area. When food is found, the strategy shifts to a more intensive examination of a small area by using slow movements and frequent turning until doing so stops being fruitful.

This foraging pattern is controlled by the reflexive and fixed reaction components of the organism's locomotor circuitry. Arms races added novel features to these systems of locomotion, and thereby enhanced area-restricted search, which remains a common basic way of finding food in complex contemporary animals, including humans.

Another key event during the Cambrian Explosion was the evolution of nervous systems with brains in bilateral animals. This provided them with novel advantages in foraging and defending, and became the basis for new arms races based on associative learning. As Simona Ginsburg and Eva Jablonka argue, one consequence of the emergence of this neural centralization was the ability to learn relationships, or associations, between stimuli, such as the taste of food and its appearance. This development, according to Ginsburg and Jablonka, had a profound effect on the course of animal evolution:

> Although the evolutionary emergence of associative learning required only small modifications in already existing memory mechanisms . . . once this type of learning appeared on the evolutionary scene, it led to extreme diversifying selection at the ecological level: it enabled animals to exploit new niches, promoted new types of relations and arms races, and led to adaptive responses that became fixed through genetic accommodation processes.

Associative learning is for the most part a process of *reinforcement learning,* in which a connection is made between stimuli with biologically significant reinforcing consequences and neutral stimuli that co-occur (sights or odors). As a result, the neutral stimuli become cues that predict the significant stimulus and elicit anticipatory innate behavioral and physiological responses that prepare the animal for acquiring a beneficial outcome (consuming nutrition) or averting a potentially harmful one (attack by a predator).

You are likely familiar with the form of associative reinforcement learning called *Pavlovian conditioning.* In his famous studies of dogs,

Ivan Pavlov discovered that playing a sound (it actually wasn't a bell) before the arrival of food led to the formation of an association between the two stimuli. That is, after this conditioning, the sound alone would cause the dogs to salivate in anticipation of the food to come.

Associative learning contrasts with *non-associative* learning, which involves a single stimulus event rather than an association between two stimuli. A prime example is habituation. For example, if the dome of a jellyfish is touched repeatedly, it habituates (stops contracting its escape muscles). Another kind of non-associative learning is sensitization. With sensitization, after strong stimulation on one part of the jellyfish body, a weak stimulus on another part of the body elicits a strong response. Unlike in associative learning, the two stimuli never occur together—the strong stimulus just primes the body to respond to the weaker one.

Ginsburg and Jablonka's suggestion that associative learning evolved by small modifications of an existing memory mechanisms was about non-associative learning. Supporting their hypothesis are findings by Nobel Laureate Eric Kandel showing that neural mechanisms underlying associative Pavlovian conditioning differ only slightly from those involved in non-associative sensitization.

Another kind of associative reinforcement process also seems to have emerged in the Cambrian period. It is called *instrumental learning.* Pavlovian associative reinforcement learning involves connecting novel stimuli to biologically significant stimuli, allowing the neutral stimuli to come to control innate responses. But in instrumental learning, the animal acquires a new (non-innate) response. The simplest kind, and the one that likely occurred during the Cambrian, is called *stimulus-response habit learning.* This occurs when environmental stimuli are connected to behavioral outcomes by reinforcers. That is, when a behavior succeeds in (is instrumental in) producing food or avoiding harm, the reinforcer is said to stamp-in an association between the neutral stimuli that are present and the new behavior. This resulted in enhanced behavioral sophistication in both invertebrates and vertebrates.

Instrumental learning, according to Hills, evolved by modifying the mechanisms of area-restricted search, specifically by fusing the area-restricted search mechanism with the mechanism of associative learning. Consistent with Hills's hypothesis is the fact that both area-restricted search and instrumental learning depend on the release of dopamine in locomotor circuits in bilateral animals. As we have seen over and over, existing mechanisms are co-opted to create new ones.

During the several hundred million years that followed the Cambrian Explosion, each phylum of animals diversified extensively, giving rise to the more than one million animal species we share the planet with today. Many of these are bilateral animals with central nervous systems that use associative learning to support their efforts to survive.

10

Vertebrates and Their Nervous Systems

That vertebrates are a distinct group of animals has long been recognized. In the eighteenth century, when biological classification systems became more sophisticated, vertebrates were assigned their own phylum—Vertebrata. But in the next century, Vertebrata was demoted to a subphylum of Chordata, which also included several invertebrate subphyla. To understand the evolutionary origins of the vertebrate nervous system, we need to understand what chordates are, and how vertebrates fit in this group.

Deuterostomes, and the Emergence of Chordates

Most invertebrates (for instance, worms, mollusks, crustaceans, insects, and spiders) are protostomes. The remaining invertebrates, and all vertebrates, are deuterostomes. The distinguishing feature of these two kinds of bilateral animal is whether the mouth or anus end of the digestive tract opens first during early life. In protostomes, the mouth wins the embryological race, while in deuterostomes it's the anus.

The common ancestor of protostomes and deuterostomes was a worm that is suitably called the protostome-deuterostome ancestor (PDA). Protostomes appeared first, diverging from the PDA roughly 600 million years ago, followed by the divergence of deuterostomes from the PDA 580 million years ago. Then, early in the Cambrian Explosion, about 540 million years ago, some deuterostome invertebrates split off from the others, resulting in the first members of the phylum Chordata. Ten million years later, fish, the first vertebrates, diverged from the invertebrate chordates.

Phylum Chordata gets its name from the *notochord,* a cartilaginous structure that runs longitudinally in the body. Early chordates were invertebrates—they lacked a vertebral column and spinal cord. Their body structure was supported by a notochord (precursor of the

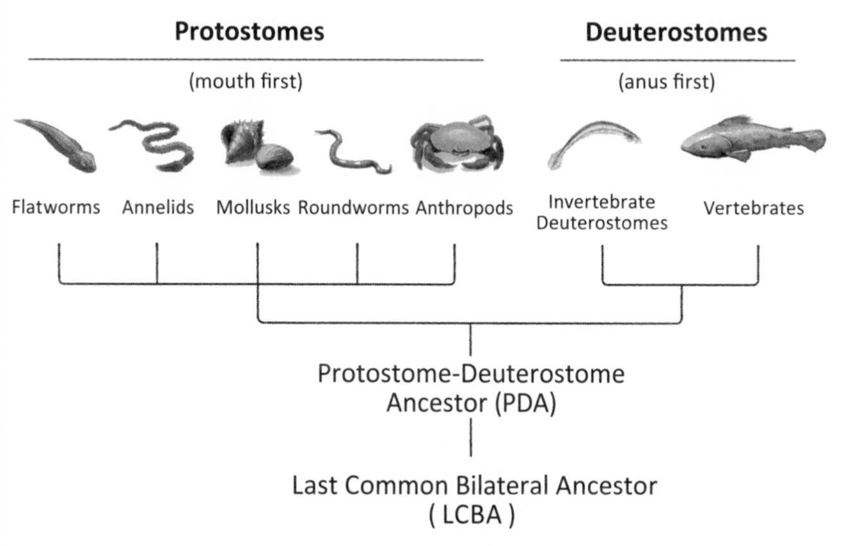

Figure 10.1. Protostomes and deuterostomes. Reformatted from Joseph LeDoux, *The Deep History of Ourselves* (New York: Viking / Penguin Random House, 2019), figures 37.1, 38.1.

vertebral column), and instead of a spinal cord they had a nerve cord. The notochord is located between the belly and the nerve cord. The brain of invertebrate chordates is an extension of the nerve cord, much like how the vertebrate brain is an extension of the spinal cord, and it provides some centralized control over the body, though less than in vertebrates. Reflecting their chordate history, all vertebrates still possess a notochord, but only during early embryological life.

Protostome invertebrates like flies also have a brain as an extension of the nerve cord. It is therefore important to consider the meaning of this similarity between vertebrates and protosome invertebrates. The brains of flies and mice don't look very similar, but nevertheless have striking parallels at the molecular and genetic levels. For example, their bodies, including their brains, are built during early development by similar families of genes and molecules, and some of the overlapping genes and molecules play a role in learning and memory.

An important example of the overlap of chemicals is the class of molecules called *cell-adhesion molecules.* In both protostomes and deuterostomes, these molecules play a key role in constructing complex bodies consisting of tissues and organs, enabling cells to "stick" together. In the nervous system, they stabilize connections between neurons following synaptic plasticity in early life, when the brain is being built. Later, they contribute to the strengthening of synaptic connections between neurons when the organism learns and stores memories.

The most likely explanation for how these genes and molecules came to be shared is that they were passed on to both groups as they diverged from their common ancestor, the PDA. Indeed, genetic evidence shows that cell-adhesion molecules and other learning-related genes shared by protostomes and deuterostomes were inherited from the PDA, which got them from Cnidaria, which acquired them from sponges, which in turn inherited them from their unicellular protozoan ancestor. This sequential set of genetic and molecular links from protozoa, which lack nervous systems, through animals with nervous systems, is another example of how existing capacities are coopted for new purposes as new kinds of organisms evolve.

The Diversification of the Vertebrates

The first vertebrates were fish that lived 530 million years ago and lacked bones, but had a cartilage-based skeleton. They were active predators. Absent a moveable jaw, they filtered food through their stationary teeth. Present-day examples are lamprey and hagfish. Fish with bones and moveable jaws, which include most of the fish we eat, came later in the Cambrian, and diversified quickly during that time.

The first non-aquatic (non-fish) vertebrates appeared about 350 million years ago. This happened when *fishapods,* which were a transition animal between fish and amphibians, adapted to a terrestrial existence by converting fins into stilt-like legs. The abundance of vegetation on land provided a food source. But as part of their respiration, plants also provided oxygen, which is a necessary ingredient for animal respiration and metabolism. Fishapods, in addition to being able to walk on all fours, also acquired other adaptations useful for terrestrial locomotion, including the ability to see on land as well as underwater. Nevertheless, they could not stray far from water since, like their fish ancestors, their eggs were fertilized externally, which required water or moist soil. Although fishapods' terrestrial vision was limited, it paved the way for distal vision in amniotes.

Amniotes (named for the amnion, an internal compartment for hosting the embryo) arrived about 310 million years ago. As a result of this reproductive change, amniotes were less dependent on a life in close proximity to water than earlier vertebrates had been, and their visual capacities on land expanded, allowing them to forage widely, and to detect prey and predators at a distance.

Reptiles evolved directly from early amniotes. A second line of descent started, too. From it *therapsids,* another amniote, evolved. Mammals, a subgroup of therapsids, diverged later.

The legs of the early amniote common ancestor of reptiles and therapsids hung from its sides. Reptiles retained this body plan, but therapsids came up with something different: legs directly underneath the body trunk. Because of this arrangement they were able to breathe and run at the same time when escaping predators or capturing prey. This was a tremendous advantage, but demanded a lot of energy, and so required them to consume more food and use more oxygen than reptiles. The resulting higher metabolic rates produced body heat, which allowed therapsids to maintain an internal body temperature through metabolism. In other words, they were endothermic (warm-blooded). And they had a visual system that allowed them to see well and forage during daytime or at night. Reptiles, being cold-blooded, had to heat their bodies externally during daylight; at night they were sessile in order to conserve heat.

A mass extinction about 250 million years ago wiped out much of animal and plant life. Dinosaurs, with their lower-energy, exothermic (cold-blooded) bodies, survived better than did the higher-energy therapsids. Even so, some small-bodied, low-energy-demanding therapsids, known as *cynodonts,* survived. Slowly, cynodonts gave up daytime foraging and the visual capacities that support it, and instead led nocturnal lives. Mammals evolved from these nocturnal cynodonts about 210 million years ago.

Early mammals, like their cynodont ancestors, had small bodies and were nocturnal. Being warm-blooded, they could spend much of their waking life in the dark, thereby avoiding predation by cold-blooded carnivorous reptiles, especially large dinosaurs, which slept at night.

Just as dinosaurs benefited from a climactic event, so did small mammals. Around 65 million years ago, another mass extinction occurred, due to climate change (likely accelerated by the impact of an asteroid and perhaps volcanic eruptions). Dinosaurs did less well this time—in fact they became extinct. With dinosaurs out of the way, small mammals were able to move more freely, invade previously avoided niches, grow in size, and ascend to the top of the food chain.

Most mammals remained nocturnal, despite little risk of predation during the day. This is believed to be because their cynodont ancestors, in the process of becoming nocturnal, sacrificed day-vision capacities, and acquired high-frequency hearing, touch-responsive whiskers, and very sensitive olfaction to compensate in the dark. Later, some mammals, especially but not exclusively primates, reacquired daylight visual capacities and came to live diurnally.

Mammals today are present on every continent, occupying terrestrial and aquatic areas in a wide variety of climates. They include the largest and smallest animals alive.

The Vertebrate Brain

Each major vertebrate transition involved selective pressures that produced dramatically different body types—a fish, frog, snake, chicken, mouse, monkey, and human have quite distinct bodies that move around in the world very differently. Not surprisingly, then, novel forms of control by the nervous system were required for these unique body plans to satisfy the universal survival needs of nutrition, hydration, respiration, thermoregulation, defense, and reproduction.

Differences in neural control in different vertebrates are best seen as variations on the same overall organizational plan of the vertebrate brain. That is, the brains of all vertebrates, regardless of which class of vertebrate they belong to, have three broad zones: a hindbrain, midbrain, and forebrain. Over the long course of vertebrate evolution, the hindbrain, involved in the control of life-sustaining visceral functions of the body, changed the least, while the forebrain, which orchestrates higher levels of behavioral and visceral control, changed the most. In particular, as each new class of vertebrate evolved, the forebrain further expanded in size and complexity.

The Three Major Zones of the Vertebrate Brain

Forebrain—learning and complex behavioral control
Midbrain—sensory-motor control over primitive behaviors
Hindbrain—control of life-sustaining body functions
(Brainstem = hindbrain and midbrain)

The mammalian forebrain, which we will be particularly concerned with going forward, consists of two main components. The highest level is the telencephalon, which is constituted by the cerebral hemispheres. These paired (that is, bilateral) structures make up much of the forebrain. Each is composed of a cerebral cortex and subcortical cerebral nuclei. By the way, all brain areas are paired in bilateral animals.

The Forebrain of Mammals

Telencephalon (cerebral hemispheres)
- Cerebral Cortex
- Subcortical Cerebral Nuclei

Diencephalon
- Thalamus
- Hypothalamus

The cerebral cortex is laminated—it consists of neurons arranged in distinct layers. And differences in the degree of lamination define three broad kinds of cortex. Cortical areas with six layers of neurons are referred to as *neocortex* (or *isocortex),* areas with four or five layers are termed *meso-cortex,* and areas with three or fewer layers are called *allocortex.* Allocortex is the oldest evolutionarily, being present in all vertebrates. Meso-cortex came next, and is especially well developed in mammals. Neocortex is the most recent, and, though present in mammals as sensory and motor cortex, it expanded considerably in primates to form multimodal or associative areas.

The subcortical cerebral nuclei have little in the way of neuronal lamination. Areas like the basal ganglia and amygdala are key examples. The basal ganglia refers to a group of nuclei connected with sensory and motor regions of the neocortex and involved in controlling behavioral responses. The amygdala receives sensory information about biologically significant stimuli that indicate food or danger or mating opportunities, and controls species-typical behavioral and visceral responses.

Examples of Key Areas of the Mammalian Cerebral Hemispheres

Cortical Areas

Neocortex (six layers):
- -sensory cortex: visual, auditory, and somatosensory cortex
- -motor cortex: motor and premotor
- -associative cortex
  - -posterior parietal
  - -superior temporal
  - -prefrontal cortex: dorsolateral, dorsomedial, ventrolateral, frontal pole

Meso-cortex (four or five layers):
- -prefrontal cortex: orbital, anterior cingulate, prelimbic, ventromedial, insula
- -temporal lobe: entorhinal, perirhinal

Allocortex (three or fewer layers):
-temporal lobe: hippocampus
-olfactory lobe: piriform cortex

Subcortical Areas

Amygdala
-basolateral nuclei
-central nuclei

Basal Ganglia
-dorsal striatum
-ventral striatum

The second major part of the forebrain is the diencephalon, which includes the thalamus and hypothalamus. The thalamus is best known for its role in transferring sensory information from the external sensory organs to the sensory areas of the cortex. A key role of the hypothalamus is to control behavioral and visceral responses of the body in order to help the body survive day to day.

Forebrain Follies: A Magical Tale about Brain Evolution

Over the years, many neuroscientists have succumbed to the temptation of treating evolution as a process by which new parts are added to old ones. This reflects the pre-Darwinian idea of life as a ladder, with each rung representing successively more evolved animals stacked on top of one another and with humans appearing on top. The ladder metaphor echoed the medieval *great chain of being,* which viewed "man" at the top of the worldly hierarchy of creatures, and so closest to God. Darwin, by contrast, proposed that evolution was more like a widely branching tree than a ladder, with humans just another branch, a side shoot of the primate twig.

The adoption of the ladder model has been particularly problematic in theories about the evolutionary changes in the forebrain over the course of vertebrate evolution. For example, around the turn of the twentieth century the German anatomist Ludwig Edinger proposed that the human forebrain sits on top of the midbrain, and reflects an assemblage of ancestral vertebrate brains stacked on top of one another—a reptilian forebrain, an early mammalian forebrain, and a new mammalian (primate) forebrain.

Edinger's line of thought was adopted by leading anatomists in the early twentieth century. But it reached a crescent of popularity, both scientifically and in the public eye, in the second half of the twentieth century through the influence of Paul MacLean's famous *limbic system* theory. (MacLean originally used the term *visceral brain* for the limbic system because of the involvement of these areas in visceral control, but the limbic system is the term that stuck.) In his model, the three stacked components of the forebrain were the basal ganglia, limbic system, and neocortex. MacLean later came to collectively refer to these three components as the *triune brain,* and assigned each with specific behavioral and mental functions based on his assumptions about which kind of behavioral control was dominant in various reptiles and mammals.

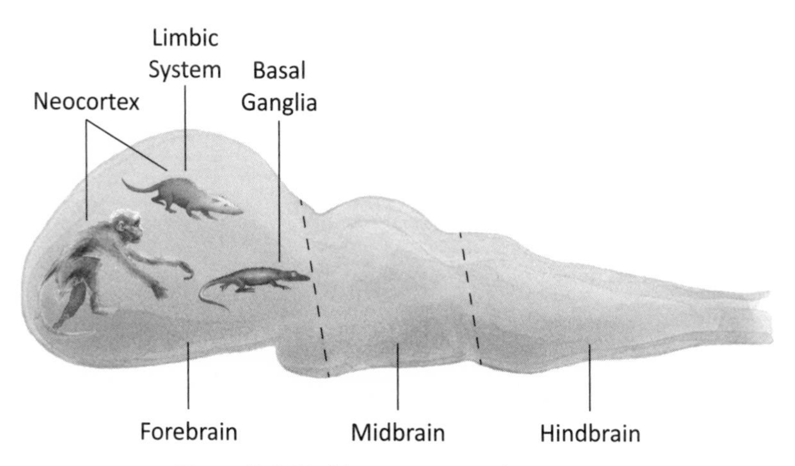

Figure 10.2. Limbic system / triune brain model

The lowest level of the triune brain was the basal ganglia, a set of subcortical motor control areas that MacLean thought were primarily involved in the control of instinctual behavior in reptiles. Mammals then, according to MacLean, evolved a limbic system that made emotions. This, he said, allowed mammals to use feelings to flexibly control behavior, freeing them from the rigid instinctual heritage of reptiles. The limbic system included subcortical forebrain areas such as the amygdala and cortical areas such as the hippocampus, cingulate gyrus, and orbital region. What tied these areas together, in his view, was connectivity with the hypothalamus. These connections, in turn, he said, made it possible for emotions to control survival behaviors and the visceral responses of the body that support those behaviors.

The limbic cortical areas are all located medially in the brains of all mammals—with the allocortex in the medial temporal lobe and the meso-cortex in the medial frontal lobe. Imagine the hemispheres of the brain as a hotdog bun. The white part, visible only when the bun is pulled apart, is analogous to the limbic cortical areas. The brown outer part of the hotdog bun is like the neocortex in the lateral parts of the hemispheres. In early mammals, the neocortex consisted mainly of the sensory and motor cortex, and was primarily used for controlling behavior in responses to sensory stimuli. These forebrain areas expanded in size and complexity in primates. In addition, primates added new areas of neocortex that engaged in the complex integration of sensory information across modalities in the temporal, parietal, and frontal lobes. This so-called multimodal *associative cortex* allowed primates to reach new heights in cognition, and in humans, language-based cognition.

Summarizing this triune brain, MacLean noted:

> Man, it appears, has inherited essentially three brains. Frugal Nature in developing her paragon threw nothing away. The oldest of his brains is basically reptilian; the second has been

> inherited from lower mammals; and the third and newest brain is a late mammalian development which reaches a pinnacle in man and gives him his unique power of symbolic language.

MacLean's use of *paragon* and *pinnacle* to describe the human brain's place in evolution are indicative of a scientific version of the ladder view of evolution. Writing about such matters, Ann Butler and William Hodos noted that organisms are neither superior nor inferior to other organisms. Through natural selection, diversity is created. As a result, when the environment changes, or the group moves to a new niche, new traits become important, and previously useful traits can become detriments. But the changes are not direct responses to the environment per se; instead, they result when the population of individuals with traits that fit the environment increases, and the population of those whose traits are less compatible declines.

In other words, we are newer and different than our mammalian and even our primate and early human ancestors, but not better. There is no goal or purpose to evolution, no pinnacle. We are not some endpoint, but instead are part of a continuous process.

The Actual Evolution of the Mammalian Forebrain

The limbic system / triune brain theory has been criticized by many neuroscientists, including me. For starters, MacLean was arguing that large anatomical structures of the brain possess distinct functions that operate somewhat independently. But this idea has been widely criticized. These days functions are much less likely to be viewed as localized in brain regions, large or small, but instead are understood to be dependent on synaptically connected networks that are distributed across areas within and between the major zones.

Contrary to the idea that mammals invented brand-new forebrain areas, compelling evidence (based on anatomical, physiological, and genetic studies) shows that pre-mammalian vertebrates possessed at least precursors or homologs of all the major forebrain regions that present-day mammals came to have. For example, the ancient jawless fishlike lamprey possesses precursors of major areas in the mammalian, including the human, brain. This suggests that the blueprint of the vertebrate brain, firmly set in Cambrian period, has been retained in all vertebrates living today.

Also contrary to MacLean's ideas, mammals did not invent limbic areas like the amygdala and hippocampus, or the cerebral cortex, or even the neocortex, and the mammalian brain was made not by stacking new parts on top of old ones, but by building on and expanding existing parts from ancient vertebrates. There were certainly additions along the way. But in most cases the changes were due to elaborations of something that already existed.

Despite the numerous problems with the limbic system / triune theory, the intuitive appeal of MacLean's ideas allowed these to survive as both cultural memes and scientific wisdom. In fact, a recent textual analysis found that ladder-like metaphors remain prominent in contemporary scientific writings about brain evolution.

Edinger and MacLean worked with what they had. Although the details of their ideas have not held up, their writings stimulated a tremendous amount of research and theorizing, including research and theories that challenged their views. With these outdated ideas put aside, the scientifically accepted view—that new features often emerge as modifications of old ones—could gain traction with a broader swatch of neuroscientists, as well as with scientists in other fields that have been influenced by the triune brain and limbic system conceptions, such as psychology, psychiatry, and philosophy. The true contributions of Edinger and MacLean could then be better appreciated, and they could be celebrated as having paved the way for, rather than having impeded, progress.

11

Romer's Rendition

Having a nervous system is what makes a neurobiological organism different from a mere biological one. But a nervous system is obviously not separate from the body in which it exists; it's a system of the body, and it is intricately entwined with other body systems. As a result, to understand how the nervous system does what it does for animals, we need to understand how it is interconnected with the body.

Redesigning the Nervous System

The traditional view of the nervous system, the one celebrated in neuroscience textbooks and the one I have discussed so far, is that it consists of two major parts: a *central nervous system* (CNS) that includes the brain and nerve cord (spinal cord in vertebrates), and a *peripheral nervous system* consisting of sensory and motor nerves that interconnect the CNS with the body.

Recall that in Chapter 8, I discussed various ideas about animal bodies having two kinds of existence—a visceral existence on the inside, and a somatic existence related to the outer world. I highlighted Romer's version of this biological duality, especially his novel

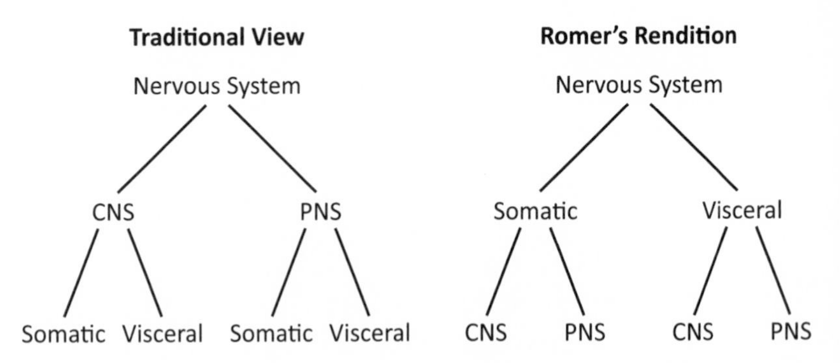

Figure 11.1. Romer's reconceptualization of the nervous system

views about how the muscles that underlie visceral and somatic functions pinpoint where the two kinds of corresponding activities take place in specific tissues of the body.

But what I intentionally left out there, so I could take up here, was the actual topic of Romer's article—how the two kinds of bodily activities are controlled by two different nervous systems. He proposed that interactions of the body with the external world are taken care of by a *somatic nervous system,* while internal bodily functions are serviced by a *visceral nervous system*.

While these visceral and somatic neural functions are usually treated as subcomponents of the central and peripheral parts of the nervous system, Romer flipped the relations around. In effect he restructured the organizational chart of the nervous system, elevating the somatic and visceral nervous systems to the level of primary partitions, and making the central and peripheral locations of their neural tissues secondary (see Figure 11.1). Why would he do this? After all, there is more to the CNS than just the control of striated muscles and visceral tissues—our brains are not just means of responding to external stimuli and maintaining homeostasis. They also think, feel, remember, and anticipate.

My Telling of Romer's Tale

My guess is that Romer was not dismissing the role of cognition in the human brain but was, instead, trying to understand the condition of early nervous systems. Based on this assumption, I will use information I discussed earlier, some of which was *not* available to Romer, to put his ideas into a broad evolutionary narrative that he might come up with if he were writing today.

Nervous systems evolved in tandem with muscles to keep primitive animals alive. For example, the diffuse (non-centralized) nervous systems of radial Cnidaria could only crudely control somatic (movement) and visceral (digestive and other metabolic) functions of the body. But the centralization that came when bilateral animals added a brain was foreshadowed by two specialized concentrations of neurons in Cnidaria, one that was involved in the capture and digestion of food and another that controlled escape behavior. When centralized nervous systems arrived, they continued in this vein, with some circuits controlling behavioral movements toward useful and away from harmful things, and others controlling inner visceral functions related to digestion and metabolism—in other words, metabolic homeostasis.

With the divergence of early bilaterals into protostomes and deuterostomes, and the divergence of vertebrates from invertebrate deuterostomes, visceral and somatic functions remained the core activities of the nervous systems—the visceral nervous system, including its central and peripheral components, managed metabolic homeostasis, and the somatic nervous system, including central and peripheral components, managed behavioral interactions with the environment. These continued to be the primary functions of the nervous system when cognition emerged in some neurobiological organisms, providing additional tools that could be used to help them stay alive and reproduce.

Romer figured out that the central and peripheral nervous systems were not the targets of natural selection. Instead the targets were those components that performed visceral and somatic functions in the body. As species diversified through evolution, circuits instantiating these functions took twists and turns over time, with some old features being retained and others giving rise to new ones. Perhaps Romer was influenced by MacLean's original designation of the limbic system as the visceral brain.

Recent studies add icing to this particular evolutionary cake, showing that the genes underlying the allocation of neurons to those circuits that control somatic versus visceral activities in the bodies of mammals come from a line of ancestry that extends back to the protostome-deuterostome ancestor (PDA), which in turn got some of the key genes from Cnidarian ancestors. That is, the somatic-visceral distinction in the nervous system is old and continuous in the history of animals.

But it didn't all start with animals. The somatic-visceral distinction exists in all organisms, including other multicellular organisms (plants

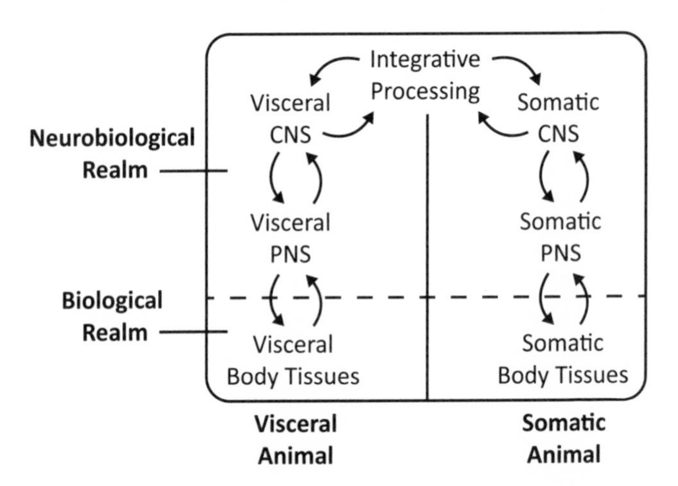

Figure 11.2. The neurobiological realm evolved from and is intricately entwined with the biological realm.

and fungi), as well as the single-cell eukaryotic ancestors of multicellular organisms, and even the prokaryotic ancestors of eukaryotes. In other words, all the way back to the beginning of life.

All organisms have to maintain metabolic homeostasis between their internal milieu and the surrounding environment, and part of how they do that is through behaviors such as approach and withdrawal. In other words, through natural selection, the visceral and somatic functions of the primordial biological realm were carried forward into the neurobiological realm as animals evolved and diversified.

To consider these two realms as entwined and interdependent is to take a quite different approach to how the brain works than that offered by the modular, triune brain model. It requires acknowledging that the partition between visceral and somatic functions was the original fault line within the nervous system from the get-go, from before centralization, and therefore from before the central/peripheral divide. Through this lens, it becomes clear that control over visceral and somatic functions of the body by the nervous system is the very reason the nervous system exists. It is, in fact, the basis of the uniqueness of animal existence, which is to say, it is the reason for the neurobiological way of being.

12

Viscerology

Let's explore visceral control in mammals in more detail. A good place to begin is with Xavier Bichat, the eighteenth-century vitalist and pioneer in tissue biology. Bichat proposed that the inner world of vegetative (visceral) functions, his *vie de nutrition,* is regulated locally by a nervous system within the body that operates independently of the brain. The brain, he said, was only concerned with the *vie de relation,* that is, with behavioral interactions with the outer world. Until recently, this was the dominant view; scientists believed that the CNS, especially the brain, had little involvement in visceral control. This bias, I believe, influenced the way we came to understand the crucial role of the brain in visceral functions.

The Visceral Periphery

The British physiologist John Newport Langley introduced the term *autonomic nervous system* (ANS) in the early twentieth century as a new designation for Bichat's vegetative nervous system. *Autonomic* was meant to capture the unconscious, non-volitional, neural processes that automatically control the visceral tissues, making homeostatic adjustments as needed, more or less independent of the higher centers of the brain.

Like Bichat, Langley noted that ANS neurons collect together as ganglia near the visceral tissues they control. The axons of these ganglia cause the tissues to respond by contraction or relaxation, or, in the case of endocrine glands, to release hormones into the bloodstream.

Langley conceived of the ANS as having two components that work in opposition. The *sympathetic nervous system,* he said, is prewired with innate discharge patterns that are sufficient to control the physiological functions of the viscera. The *parasympathetic system,* by contrast, was said to operate alongside the sympathetic system (hence the designation *para*), and counteract the sympathetic system's effects and restore equilibrium. Both systems were believed to affect all visceral organs and glands in a singular, all or nothing, reflexive fashion.

A key feature of the somatic nervous system is that it has sensory and motor components. For example, if you touch a hot stove, sensory messages travel to the spinal cord and automatically cause your hand to withdraw. But ANS ganglia, especially the sympathetic ganglia, control visceral responses without the benefit of sensations from the viscera. For this reason, Langley viewed the ANS as a motor system.

While Langley's terminology is still used today, our contemporary understanding has led to several major revisions. For example, it is no longer accepted that the two components each respond in an all-or-nothing way, affecting all visceral tissues similarly. Instead, it is thought that different situations elicit subtly different patterns of activation of the visceral organs by each part of the ANS. And the parasympathetic component, which was often treated as secondary to the sympathetic system, is no longer seen as a mere counterforce to sympathetic responses. Moreover, the ANS is no longer viewed as simply a motor system. Instead, visceral tissues respond to mechanical pressure and chemical stimuli (oxygen or carbon dioxide). Nerve endings in visceral organs even transmit viscerosensory messages to the hindbrain via the parasympathetic vagus nerve.

Some researchers feel that these tweaks are futile efforts to salvage a failed idea. The neurobiologist William Blessing, for example, has forcefully argued that the concept of the ANS has outlived its usefulness. Nevertheless, the ANS, like the limbic system and triune brain, lives on, despite such scientific challenges.

While the sympathetic and parasympathetic components of the ANS have historically received most of the attention, a third part also exists. This *enteric nervous system* consists of millions of neurons that are associated with the digestive tract, but also with cardiac and some other visceral tissues. In the late twentieth century it gained scientific attention, and was nicknamed the *gut brain,* which garnered it a place in the public imagination. Unlike in the other two parts of the ANS, enteric neurons are embedded in the visceral tissues themselves rather than in external ganglia.

The Visceral Brain

Langley's idea that the brain is minimally involved in ANS control was wrong for three reasons. First, as we just saw, the parasympathetic vagus nerve sends signals to the hindbrain about the state of the body. Langley knew this but downplayed it. Second, visceral sensations in the hindbrain clearly connect with the midbrain and forebrain. As early as 1884, William James had proposed that visceral sensations are crucial to the conscious experience of emotions, which would require forebrain reception of signals from the viscera. Either Langley did not know about James's view or he dismissed it.

Another way that Langley was wrong about the ANS and the brain was in his equation of forebrain function within consciousness, including the volitional control of behavior. His logic: because control of the ANS is automatic (unconscious and non-volitional), it must take place totally in the periphery and brainstem, not the forebrain. But scientists today routinely accept that many aspects of cognitive processing by the forebrain take place non-consciously, with only a fraction of this processing reaching conscious awareness. In other words, just because the ANS operates unconsciously doesn't mean that the forebrain is not involved in its control.

The involvement of the forebrain in ANS functions had actually begun to come to light in Langley's time. During the early twentieth century, German researchers reported that electrical stimulation of the subcortical forebrain, especially the hypothalamus, elicited changes in cardiovascular and other visceral activities. Around the same time, Walter Cannon, of homeostasis fame, discovered that adrenaline and noradrenaline released from the adrenal medulla during fight-flight and other emergency reactions was due to hypothalamic activation of the ANS. Cannon became a leading authority on the brain's control over the viscera, and through his research, the hypothalamus, contrary to Langley's view, came to be known as the *head ganglion* (like a CEO) of the ANS. But in assuming that adrenaline maximally activates all viscera to the same degree, Cannon also gave support to this aspect of Langley's view.

With improved techniques for studying physiological responses controlled by the ANS in awake, behaving animals, later researchers inspired by Cannon showed that electrical stimulation of the hypothalamus elicited specific patterns of visceral responses associated with defensive and aggressive behaviors. The ability of electrical stimulation of the hypothalamus to coordinate behavioral and ANS responses illustrates a key way that somatic and visceral activities are integrated.

Modern research has shown that the hypothalamus influences the viscera by way of connections with those midbrain and hindbrain areas that house sympathetic and parasympathetic control circuits. Through these top-down connections to the peripheral ANS nerves and ganglia, key physiological processes underlying the homeostatic control of digestion and energy management, fluid and electrolyte balance, respiration, cardiac and vascular activities, blood gasses and pH, core body temperature, and more, are regulated.

The hypothalamus is not the only forebrain area involved in visceral control. Many of the inputs to the hypothalamus relevant to visceral control come from the limbic region of the forebrain (recall the distinction between the "limbic forebrain" anatomical concept and the problematic "limbic system" functional concept). More recent work has shown that the amygdala and some other limbic regions (including several relatively primitive areas of medial mesocortical prefrontal cortex, such as the prelimbic, orbital, anterior cingulate and insula cortex) also have direct connections with midbrain and hindbrain areas that control sympathetic and parasympathetic activities.

It's worth pausing here to note that subcortical brain areas like the amygdala, hypothalamus, and basal ganglia have subdivisions that have functional roles. For example, the lateral region of the amygdala in mammals receives sensory inputs, while the central amygdala area connects with visceral control areas of the hypothalamus and brainstem. Because these two parts are interconnected, a loud sound that activates the lateral amygdala can result in increases in heart rate and blood pressure by way of the central amygdala.

The enteric nervous system is the closest of the three ANS components to being independent of the CNS, providing some vindication for Bichat and Langley. But even it is subject to CNS oversight. A key level of control involves parasympathetic circuits in the hindbrain that connect with the enteric system, and that are regulated by visceral forebrain areas. In other words, the so-called gut brain is also not completely free of the real brain.

Visceral Loops

The brain not only acts on the viscera; it also receives signals about the status of visceral tissues. Recall that visceral tissues have sensory nerve endings that terminate in the hindbrain. Hindbrain circuits then connect with midbrain circuits, which in turn connect with the forebrain, especially the hypothalamus and amygdala, but also with meso-cortical prefrontal areas such as the orbital, anterior cingulate, and insula cortex. Each of these levels of the brain both receives inputs from the level below it and sends inputs back to the lower level, in a network of information loops.

The endocrine system mediates other interactions between the viscera and the brain. As noted earlier, the hypothalamus, via the ANS,

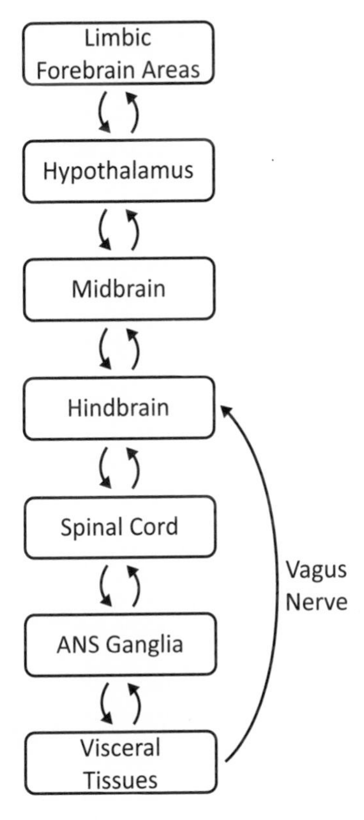

Figure 12.1. Visceral loops

directs the adrenal medulla to release adrenaline and noradrenaline in stressful situations (recall Cannon's emergency reactions). These hormone molecules are too large to enter the brain from the bloodstream, but they affect the brain by binding to receptors on the parasympathetic nerves in the chest and abdomen. The vagus nerves, as we've seen, then send messages about the state of the viscera to hindbrain areas that, in turn, connect with midbrain and forebrain areas. Also in stressful situations, the hypothalamus directs the pituitary gland to release adrenocorticotropic hormone (ACTH), which travels in the bloodstream to the adrenal cortex, resulting in the release of the hormone cortisol. Unlike the hormones of the adrenal medulla, cortisol is small enough to enter the brain from the bloodstream, then bind to steroid receptors on neurons in the forebrain.

Visceral States

I have discussed the relation between the brain, ANS, and viscera as if there is a single visceral brain system involved. This is the tradition handed down from Cannon and MacLean. But we now know that there are specific circuits, rather than centers, that control different kinds of visceral activities. For example, energy homeostasis, fluid and electrolyte balance, temperature regulation, and reproduction each depend on different circuits with components centered on the limbic forebrain, hypothalamus, and brainstem. Moreover, visceral control by circuits in these areas participate in processing loops between visceral tissues in the body and the various levels of the CNS. For example, when energy supplies are low, signals from the digestive system in the body are transmitted to the brain via both the vagus nerves and hormones. Top-down control by the energy management circuit in the forebrain directs the ANS to release energy stored in the liver or in fat cells, and initiates behaviors to acquire new supplies.

Visceral states non-consciously contribute to body homeostasis. But non-conscious does not mean non-consequential. These states reflect the body and brain working together to support our well-being.

Arousal

Another important set of brain circuits involved in homeostasis is located in the core of the brainstem (that is, the midbrain and hind-brain). In the past, these circuits were referred to as the *reticular formation,* which was thought to be a unified global arousal system, a network of neurons that regulates the basal-level activity of the brain. Modern understanding emphasizes collections of neurons that make specific chemicals and release these widely throughout the brain. Included are norepinephrine, serotonin, acetylcholine, dopa-mine, and a relatively new entry to this list, orexin, a chemical re-leased by hypothalamic cells.

These general-purpose chemicals are said to regulate brain and body arousal levels, mobilizing or energizing the organism in a general way, and supporting activities required to restore homeo-stasis. Although released widely, these arousal chemicals have their effects by regulating neurons that are active in the moment. Hence, when defensive, energy-management, reproductive, or other circuits are active, the arousal chemicals affect their processing more than when they are inactive. This is why these chemicals are called *neuromodulators*—they regulate rather than initiate neural activity.

A particularly important role of arousal systems is to regulate sleep-wake cycles. You may not realize it, but sleep and wakeful-ness are viscerally related functions that involve the whole organism. And the transition between these states is homeostatically regulated by physiological processes in both the brain and body.

Plasticity of Visceral Reflexes

While visceral reflexes are wired by evolution to respond to certain classes of stimuli, they can, through learning, come to be controlled by environmental stimuli. Recall the two kinds of non-associative learning processes discussed previously—habituation and sensitization. A loud noise can make your heart beat faster, but with repetition you will become used to the noise and have a milder response. A strong stimulus like a loud noise or an electric shock can also sensitize you, causing your heart to beat faster to other stimuli present in the moment that that normally would not affect you.

But visceral reflexes can also be conditioned associatively, as in Pavlov's conditioning of dogs to salivate when they hear the sound that usually comes before a treat. This kind of learning depends on the convergence in the brain of sensory circuits that process the neutral sound and the food reinforcer.

Surprisingly, despite how widely known Pavlov's study is, the anatomical details of how it works in the brain are not well understood. This is most likely because saliva is difficult to collect and precisely measure in animals that are being conditioned and tested. But other kinds of conditioned visceral reflexes have been studied more extensively. An especially well-researched example involves a procedure called Pavlovian threat conditioning, in which neutral stimuli followed by a painful event become a warning signal of danger—and come to control in a coordinated way the visceral and behavioral response to that stimulus—even when the painful event is absent.

Can Visceral Tissues Be Voluntarily Controlled?

Since ancient times the viscera have generally been assumed to be outside of voluntary control, a point of view reinforced by Bichat and Langley. There was a period in the 1960s and 1970s, however, when researchers tried to use operant conditioning and biofeedback to reinforce visceral changes, especially cardiovascular changes, in the hope of providing clinical tools that could help manage cardiovascular problems related to stress. A review of this research in 1973 concluded that while statistically significant results had been obtained in some studies, the effects were small and fleeting. The authors presciently noted that alternative modes of producing changes in visceral responses, such as training in progressive relaxation, meditation, or various yoga exercises, may prove to be more fruitful. Indeed, relaxation training, deep breathing exercises, and mindfulness meditation are common tools in psychotherapy today.

13

The Behavioral Thoroughfare

Sir Charles Scott Sherrington was an early pioneer of nervous system research. He received the 1932 Nobel Prize for his work on the neural control of skeletal reflexes, and he was the first to identify visceral interoceptors. Sherrington once mused, "The brain seems a thoroughfare for nerve-action passing its way to the motor animal." The thoroughfare to which he was referring is the flow of information from sensory systems that process external stimuli to motor systems that control muscles and move body parts, all of which allows the organism to act on and in the world. This flow of information from sensation to movement underlies behavior, and the control of behavior is a primary job of the somatic nervous system.

The external nature of this process led scientists in the late nineteenth and early twentieth centuries to use a simple strategy for pursuing the neural basis of behavior: (1) follow the peripheral nerves from sensory organs (like the eyes or ears) into the central nervous system, or CNS; (2) follow the peripheral motor nerves out of the CNS effector organs (the skeletal muscles); and (3) identify the neural "dots" that connect the sensory systems to the motor systems.

Sensing

The world in which we live is known to us through the sensory capacities of our nervous system. When we marvel at the beauty of the sky at sunset, chemical processes in our eyes are converting wavelengths of electromagnetic energy into neural signals that let us perceive the technicolored pattern playing out in front of us. The underlying sensory elements, which we hardly notice, if at all, are essential, albeit nonconscious, components of the perceived image. The image is infused with meaning by memories of similar patterns of sensations in the past, which allows us to experience those patterns as a "sunset."

While the full human experience of the sunset depends on memory and cognition, even animals that lack our degree of complex cognition, or that lack cognition altogether, can respond to the sensory cues that constitute our notion of sunset. Nocturnal animals, for example, are sensitive to the changes in luminance as the sun sets. This is what allows them to become more active as daylight dwindles. Crickets begin chirping around that time. Other animals, like wood lice, stop moving as the darkness unfolds. As Sara Shettleworth, a prominent behavioral biologist, pointed out, such behaviors are reflexive responses to the level of illumination, and are not due to any cognitive representation or conceptual understanding of darkness per se.

In his 1934 book *A Foray into the Worlds of Animals and Humans,* the Estonian scientist Jakob von Uexküll invited readers to imagine a tick waiting on a leaf to suck the blood of a host. When the tick detects butyric acid, a chemical given off by mammals, it leaps. Upon hitting the warm surface of the host, it finds a target spot (usually a hairless one) to bore in and extract blood. The tick has no concept of what it is doing. It just has simple, motor responses that are innately wired to the chemical stimulus. And that's all it needs to survive in its environment.

Von Uexküll proposed that every species experiences the sensory world in a unique way. He called this special sensory universe an animal's *umwelt*. His point was that that there is no single objective sensory reality; the world out there is different for each kind of organism, and it can be understood only in terms of the experiential capacities of the individual organism, whether it is a single cell or a complex animal.

Sensory processing starts with sensory detectors, specialized receptors that are "keyed" to the particular stimulus qualities of each sensory modality. In animals with a CNS, the receptors do an initial analysis and then send the results, via peripheral sensory nerves, to the sensory system in the animal's brain, which constructs its sensory "reality," its umwelt, in a species-typical way.

This is possible because the basic wiring of an individual's brain follows a species-typical genetic brain plan. While an animal is still in the womb, long before explicit interactions with the actual sensory world routinely take place, genetically controlled patterns of spontaneous neural activity begin to shape sensory connectivity in a way that is compatible with the sensory environment in which its species evolved. As sensory connections develop, they begin to transmit some sensation prenatally (for example, sounds from the outer world).

An animal's experiences with its sensory world after birth are also a crucial factor in creating the synaptic connections that make its sensory systems function properly. For example, studies conducted in the 1960s by David Hubel and Torsten Wiesel showed that surgical closure of one eye of an animal shortly after birth prevented cells in the visual areas in the neocortex from having the proper inputs from that eye. This led to problems in depth perception, which requires integration of inputs from both eyes. If the eye was surgically closed during a narrow time period early in life, the so-called critical period, the effects were lifelong, whereas if the eye was closed after that period, the effects were only temporary. A more extreme illustration involves people born blind or deaf who develop especially acute capacities in other sensory systems, because the stimuli that activate the functioning sensory systems lead to the development of neural connections in the unused synaptic space.

The natural world is full of examples of how experiences with sensory stimuli refine sensory processing. The olfactory system of young flies is "tuned" to its early olfactory environment. Young birds learn the subtleties of their kind of song by listening to adults sing. People living in artic areas learn to recognize the shapes of different kinds of snowflakes. A person growing up in Japan and without exposure to English will likely find it difficult, as an adult, to hear the difference in American speech between the letters R and L.

Much of sensory learning involves repeated experiences with similar kinds of stimuli. You know the color red not because red is an actual category of something in nature, but because your visual system has experienced many exemplars in the spectral ballpark that you have come to know in relation to the word *red*. Through this learning, you acquire expectancies about what red is based on the various things you have come to associate with this spectral range. You never have to affirm that a red apple is red, even though the actual color of "red apples" is quite variable.

Repeated experiences of this type help a sensory system become more responsive to, and discriminate between, specific stimuli. For example, the visual system of an experienced fungi forager can detect subtle differences in the appearance of various mushrooms, which helps them avoid a potentially harmful mistake.

One way that sensory systems learn is by accumulating information over the course of one's life. This is a form of deep learning called the *statistical learning of sensory regularities,* and it occurs in both somatic and visceral systems. With this kind of reinforcement learning, the reinforcer is not food or biological stimuli, but simply the confirmation of a prediction based on past experience—though the prediction is not a conscious one made by the psychological organism, but instead a mindless prediction made by its neural circuits. When such expectations are violated, a *prediction error* occurs, and this triggers new automatic learning.

Sensory systems also learn by forming Pavlovian associations between sensory stimuli and biological reinforcements (such as food or pain) or psychological ones (like social approval). For example, neurons in the auditory cortex tend to be responsive to tones in a certain range of sound frequencies. If a tone in one of those frequencies is paired with an electric shock, the auditory neurons respond more strongly to tones at or close to that frequency. This enhanced "tuning" to a sound that predicts harm makes for a brisker sensory response, which in turn allows neurons in downstream areas such as the amygdala, hypothalamus, and brainstem to more efficiently control the behavioral and visceral responses that result.

Behaving

In the simplest sense, behavior occurs when processes within an organism cause it to move. It is common to think of such movements as serving psychological goals. But, as I argued in *The Deep History of Ourselves,* behavior did not arise for psychological reasons. It is as old as life, and is simply part of the organism's survival toolkit. Behavior exists to keep the organism alive. This is true for all organisms, regardless of whether they possess a nervous system. Even those neurobiological organisms with cognitive and conscious capacities maintain the ability to respond non-cognitively and non-consciously.

The hallmark of behaviors that typify the mere neurobiological realm is that they are inflexible reactions. When organisms express these behaviors, they are responding automatically—given the stimulus, the response consistently occurs. We've already encountered the three general categories of these reactions: reflexes, species-typical reaction patterns, and habits. The neural circuits underlying these are remarkably conserved across the full range of vertebrates, spanning fish, amphibians, reptiles, birds, and mammals, including humans. There are, of course, differences in complexity, but they are mostly variations on a common theme.

Reflexes

Reflexes, as noted earlier, are innate motor programs that are genetically preordained to be activated by specific stimuli. These motor programs are mostly built into circuits in the spinal cord, and / or the brainstem. Peripheral sensory nerves bring sensory information into the CNS motor program from peripheral receptors, and peripheral motor nerves connect the CNS circuits to the muscles. Although no behavior is strictly inborn or strictly due to learning, some, like reflexes, are more innate than others. As with visceral reflexes, somatic reflexes can be modulated by higher levels of control in the forebrain.

A given reflex typically engages a limited set of specific muscles. But in complex behaviors, multiple reflexes are often engaged simultaneously. For example, walking is a voluntary, often goal-directed, activity, but underlying the act itself are numerous reflexes that control the coordinated movement of one's two legs and arms, all while maintaining balance and posture. The basic coordination processes begin to unfold in early life under the direction of a developmental genetic program, but much trial-and-error learning is also required for the activity to become second nature. Once learned, movements like this, which involve precisely timed contractions and relaxations of many muscles throughout the body, depend on circuits in the cerebellum region of the hindbrain. We know this in part because when the cerebellum is damaged, the person suffers devastating losses of control over their movements.

Although animals are born prewired with basic stimulus-response reflex connections, other reflexes can be acquired through Pavlovian associative learning. I mentioned Pavlov's famous experiments on conditioned salivary reflexes in Chapter 12. But Pavlov's method has also been used in connection with somatic reflexes. For example, limb withdrawal can be conditioned by pairing a sound with a shock delivered to the limb, and eyeblink can be conditioned with a sound paired with a shock or puff of air to the eyelids. Once the pairing occurs, the sound alone will elicit the reflex. The sound and shock sensory pathways converge in the cerebellum, which is the key site for associative learning of these reflexes by the brain. But neurons in the sensory pathways that are antecedent to the cerebellum also undergo conditioning, and the resulting learning facilitates the execution of the motor response by the cerebellum.

Species-Typical Action Patterns

Species-typical reactions evolved by adding more complex behaviors to the basic neural mechanisms of reflexes that are common to all vertebrates. Reflexes and species-typical reaction patterns are, in effect, interrelated components of the innate survival repertoire of the species. For example, when food is acquired by species-typical hunting behaviors, reflexive chewing and swallowing take over, then deliver the food to the lower digestive tract so it can extract nutrients.

The universal survival requirements of living things are, as I noted earlier, incorporation of nutrients, balancing of fluids and electrolytes, avoiding harm, and reproducing. Animals achieve these by eating, drinking, defending, and mating. But the way they respond is specific to their kind. For example, different species seek different things as food and treat different things as dangerous, and how they respond depends on the kind of body they possess. In general, species-typical sights, sounds, and/or odors that represent foods, liquids, or other needed substances elicit species-typical approach behaviors, while stimuli related to predators and other harmful conditions elicit species-typical defensive behaviors such as freezing, flight, or as a last resort, fighting.

Humans rely more on learning and culture than on species-typical responses, but we do have some innate tendencies or predispositions, such as a propensity to respond in a certain way to heights, snakes, seeing an aggressive person nearby, the sound of a crying baby, and sexual stimuli. We can easily tell a possible evolutionary story about such stimuli. Because they were particularly dangerous to our primate ancestors, snakes, spiders, and heights are common triggers of phobic anxiety in people. Responding to stimuli associated with reproduction, by contrast, helps our species persist, but can also be problematic, such as when one is hyperaroused or has a sexual addiction or stalking disorder.

Survival behaviors depend on *survival circuits.* Much is known about the survival circuits that underly defensive, feeding, drinking, and sexual behaviors. When a particular survival circuit is activated, a *global organismic state* of a particular type occurs. We consciously experience these as feelings of fear, hunger, thirst, or sexual desire, but neurobiologically they merely exist as non-conscious states that contribute to particular survival situations. But non-conscious does not mean non-consequential. The states reflect the integration of brain and body activities that help keep the organism alive and well.

A survival circuit that I have investigated for many years is the amygdala-centric circuit that controls defensive behaviors and supportive visceral responses in the presence of innate and Pavlovian conditioned threats. I prefer the expression *defensive survival circuit* over the more traditional notion that the amygdala is fear center. In my view, the circuit does not produce fear. It detects and responds to danger. For the same reason, I use the expression *Pavlovian threat conditioning* instead of the more traditional expression *Pavlovian fear conditioning.* I will elaborate on these ideas in Part V.

Though seemingly unnatural, Pavlovian threat conditioning is an excellent way to model danger in the wild. The tone-conditioned stimulus mimics sounds that might occur in a real encounter with a predator, such as the crackling of leaves as a predator is about to pounce, and the shock unconditioned stimulus mimics the painful sensation of being wounded by a predator. In future encounters with the conditioned stimulus, defensive behaviors and supporting cardiovascular changes are elicited before the dangerous stimulus causes harm.

Major components of the defensive survival circuit, as revealed in rodent studies (and confirmed in lower vertebrates, birds, monkeys, and humans) are depicted in Figure 13.1. Although the amygdala is a key component, the overall circuit involves much more than the amygdala. Species-typical threat stimuli are transmitted from sensory receptors in the eyes, ears, or nose to sensory areas of the thalamus and neocortex. These then connect with the amygdala circuit. The lateral amygdala, the sensory gateway, detects the danger signified by the threatening stimulus and connects with the central amygdala. The central amygdala then activates hypothalamic and brainstem networks that separately unleash both the muscle contractions that constitute defensive behavior, especially freezing, and the supporting visceral responses. The visceral responses provide metabolic support to the defensive behaviors, both in preparation for and during the response, and to restore homeostasis when the threat level decreases. If the animal is wounded during the encounter, visceral activities facilitate recuperation and healing. While the amygdala is crucial for threat conditioning, plasticity in other forebrain areas, including sensory areas, also influences the efficiency with which the defensive behavior is controlled by the defensive survival circuit. In addition, though not shown in the figure, areas of the mesocortical prefrontal cortex and the hippocampus also contribute.

All vertebrates have an amygdala and use its defensive survival functions to cope with danger. According to recent findings, the amygdala is an invertebrate chordate invention that vertebrates inherited. Although protostome invertebrates like flies do not have an amygdala, they have their own defensive survival circuits for dealing with the dangers in their lives. Defensive survival functions also exist in unicellular organisms, even though they do not have a nervous system and hence lack survival circuits.

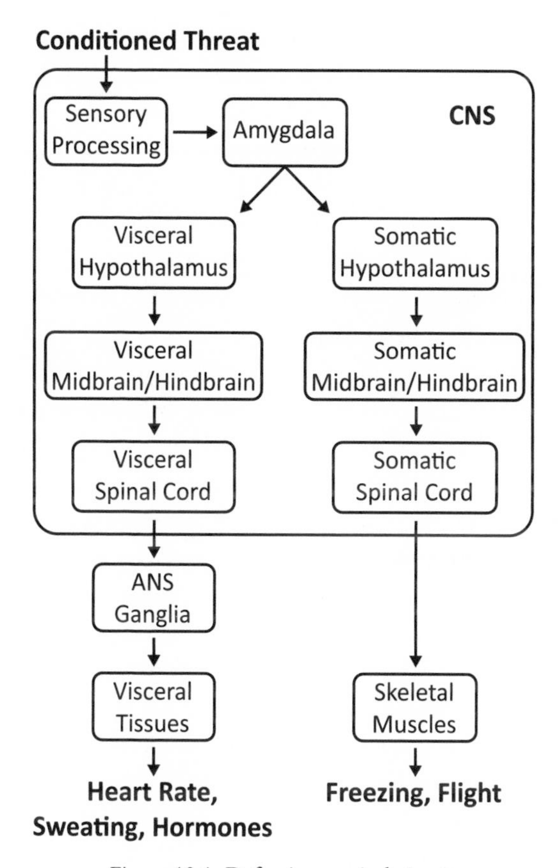

Figure 13.1. Defensive survival circuit

Habits

Animals, including people, not only learn Pavlovian associations between neural and reinforcing stimuli (stimulus-stimulus associations), but also can acquire novel behaviors by forming associations between stimuli and responses (stimulus-response associations). This form of reinforcement learning, as we have seen, is called instrumental habit conditioning.

In the late nineteenth century, the psychologist Edward Thorndike placed a food-deprived cat inside a test box where it could see and smell a piece of fish but could not reach it. The cat became very animated, making random movements. Over time the cat learned, through trial and error, that when it behaved in a particular way, the door would open. From that point on, the animal repeated the behavior when it was again in that situation. Thorndike called these instrumental behaviors because they achieved something for the organism—they provided some degree of mastery over an environmental situation.

Like Pavlovian associative conditioning of reflexes and species-typical action patterns, habits are automatically elicited by stimuli. But the two processes also differ. Pavlovian conditioning is about how a reinforcer (like food) enables a novel stimulus (for example, a tone) to control an innate response (to approach and consume the food). Habit learning, instead, involves acquiring a new behavior (such as learning how to open a locked door so that food can be approached and consumed). And in terms of learning, Pavlovian conditioning involves stimulus-stimulus associations (between a neutral sensory stimulus and a reinforcer), while in habits the reinforcer establishes stimulus-response associations.

The stamping-in of a stimulus-response habit requires integration in the brain of information about the stimulus situation with information about the response. In all vertebrates, from fish to humans, the key stimulus-response convergence takes place in the basal ganglia, which, as I noted earlier, is a collection of subcortical cerebral nuclei that play an important role in the control of movements.

The basal ganglia are reciprocally connected with cortical areas, including the sensory and motor cortex, which provide it with stimulus and response information. When the animal tastes a reinforcer, dopamine neurons in the midbrain are activated, resulting in dopamine release in the basal ganglia, and stamping-in the connection (association) between immediately present stimuli and immediately performed responses. The key part of the basal ganglia is its dorsal lateral region, and especially the dorsal lateral striatum. (This detail is important because later in the book we will see that other subregions of the basal ganglia contribute to goal-directed instrumental learning, which, unlike habit learning, requires cognitive processes.)

Once the stimulus-response habit association is acquired, the presence of the stimulus automatically elicits the response. Involved are cortical sensory-motor connection loops with the dorsal lateral striatum, and outputs of these to midbrain and hindbrain motor areas. Connections with spinal motor circuits then allow peripheral motor nerves to control the muscle contractions that underlie the habitual behavior.

Given this, we can speculate about what went on when Thorndike's food-deprived cat stumbled on a response that opened the cage door, providing access to food. At that point, the taste of the food likely caused dopamine to be released in the dorsal lateral

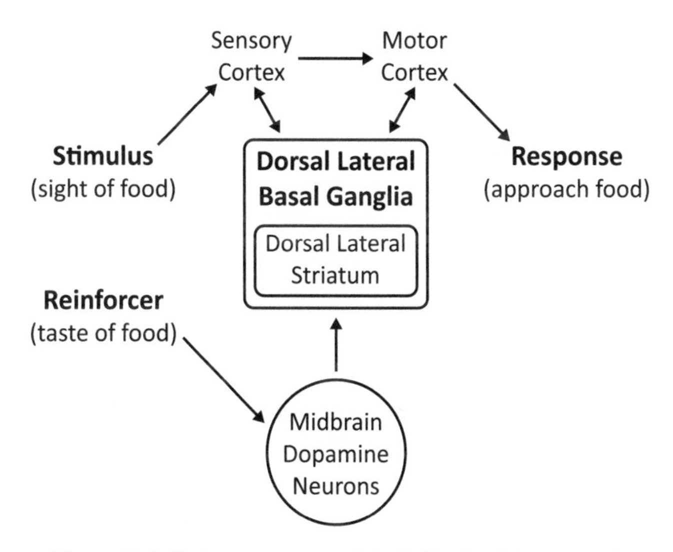

Figure 13.2. Basic components of the habit circuit in mammals

striatum, stamping-in the habit. When the cat was again in that (or a similar) stimulus situation, the habitual response was automatically elicited.

As mentioned earlier, it has been proposed that instrumental learning evolved in the Cambrian era by modifying the dopamine regulation of locomotor circuits in area-restricted foraging, and by integrating this mechanism with associative learning. Indeed, it is known that animals learn habitual responses as they move around in search of food. When food is found in a particular location, stimuli that identify that location are stamped in by the payoff. But being able to find that location again requires some understanding of the relationship of that location to other locations in the foraging area.

Although MacLean had made the hippocampus the centerpiece of his emotional limbic system, these days the hippocampus is known for its role in creating spatial maps of the environment, remembering them, and using them in foraging. In particular, the hippocampus is connected with sensory areas of the neocortex and the dorsal lateral striatal habit circuit, and, from vertebrates' earliest days, these areas have been involved in stimulus-response learning during foraging.

Habits may seem simple, but they are not. Leading habit researchers, such as Bernard Balleine, Trevor Robbins, Barry Everitt, and Ann Graybiel have argued that the final behavior that results from habit learning is a concatenation or grouping of diverse component responses, so-called action chunks, that reflect the behavioral history of trial-and-error learning that stamped in the habit.

Somewhat like learned skills (typing, gymnastics, playing a musical instrument), habits are motor responses that are acquired through repetition, and that impart mastery. The chunked action sequences of habits in fact are a very important part of learned skills. But there is a key difference. A skill is a kind of expertise that one typically acquires intentionally, rather than automatically, and then continues to work on in order to maintain or improve performance.

Habits are very useful in life since they free the brain from having to make choices in recurring situations. But they have drawbacks. To the extent that the organism always responds more or less the same way, habits are rigid, which is one of the reasons they are now distinguished from goal-directed behaviors, which are flexible. Habits, additionally, have a dark side—bad habits, like biting your fingernails or the compulsive taking of addictive drugs, are very hard to break.

The Link between Behavior and Psychology

The reflexes, species-typical reaction patterns, and habits of the neurobiological realm are often studied by psychologists, and thus considered psychological in nature. I maintain, however, that just because these are behaviors does not mean they are psychological behaviors. They are instead neurobiological realm responses, and hence involve neither cognition nor consciousness. At the same time, they are fair game for, and are in fact crucial for, psychological science since they are benchmarks above which behaviors must rise in order to be considered subject to cognitive or conscious control.

This takes us to the end of our exploration of the codependent, entwined, conjoined twins of neurobiological existence—the visceral and somatic nervous systems that sustain biological life in animals. Behavior always requires the metabolic support of the viscera, but the viscera also require behavior to supply the energy for metabolism.

PART IV

THE COGNITIVE REALM

14

Internalizing the External World

If we define biological existence the way I have—as a property of organisms—it's relatively easy to separate biological beings from non-living matter. All organisms, and only organisms, exist biologically. If something is alive, it is an organism. And if it is an organism, it is alive. While scientists disagree about whether entities like viruses should count as living things, this debate has little bearing here, since the biological beings I am mainly concerned with are also neurobiological beings, and viruses are far removed from this evolutionary history.

Deciding which biological beings exist neurobiologically is also relatively straightforward. If an organism has a nervous system, it is a neurobiological being and, by definition, it exists neurobiologically. That, in effect, narrows things down to animals, or at least most animals.

When it comes to the cognitive realm, things are considerably more complex. Since there are no physical properties that can be called on to decide which animals exist cognitively, the workings of the cognitive mode of existence must be inferred from other properties, most typically from behavior. And that can be tricky because different scientists define cognition differently. But I embrace a fairly common criterion for distinguishing cognitive from non-cognitive behavioral control—namely, the ability to use the internal representation of information to construct mental models of the world.

From Darwinian Mentalism to Stimulus-Response Psychology

Cognition in the form of thinking and knowing had been part of discussions about the mind since at least ancient Greece. But the contemporary scientific study of cognition developed from events spurred on by the Darwinian revolution in biology. Darwin's proposal that continuity between animals and humans is not limited to bodies, but also includes minds, greatly affected the young field of scientific psychology that was emerging in the late nineteenth century, especially the study of animal behavior by comparative psychologists. Following Darwin's lead, these psychologists assumed that since humans have mental states that control voluntary actions, other animals may as well.

A key goal of comparative psychology, then, was to uncover evolutionary forerunners of human mental life. To do this, they focused on so-called intelligent behavior in animals, studying their problem-solving abilities. Darwin's acolyte, George Romanes, a leading comparative psychologist, said what many were thinking—that behavior is an ambassador of the mind. And the more complex the behavior, the more likely it was a readout of the animal's mind.

In searching for animal intelligence, comparative psychologists adopted Edward Thorndike's instrumental habit-learning paradigm. Thorndike and other animal behaviorists of the time assumed that the reinforcers in instrumental conditioning tasks select the behavioral traits of individuals much like natural selection determines the biological traits of a species.

Early on he wrote about habits as being stamped in by mental states of pleasure or aversion, which fit with the anthropomorphic sentiments of the day. But a bit later, as behaviorism was kicking in, he changed his tune, saying that just as there is no intentional agent in evolution, instrumental behavior is selected (stamped in) without an intentional agent being involved.

John Watson was moved to found behaviorism because of the kind of unchecked enthusiasm for mental states as causes of behavior that Thorndike initially espoused. Building on Darwin's notion of the continuity between humans and other animals, Watson argued that animals are suitable for learning about human behavior. His initial position was that all behavior could be explained by chains of Pavlovian-conditioned reflexes. But Thorndike's instrumental habit paradigm quickly became the dominant research tool because, unlike Pavlovian conditioning, it results in the acquisition of novel behaviors. This empowered behaviorists with a sense that studies of instrumental behavior in animals could indeed provide universal laws to explain complex human behaviors, including speech, in terms of chains of stimulus-response connections stamped in by reinforcers.

The Resistance

The behaviorist stimulus-response idea cast its spell primarily over American psychologists. In Germany, Gestalt psychologists dismissed outright the stimulus-response approach. They treated organisms as entities that create meaning by organizing and interpreting their world. A leading Gestalt psychologist, Wolfgang Kohler, observed that chimpanzees, when faced with a complex problem such as how to reach inaccessible food, would suddenly come up with a solution—say, using a stick to bring the food closer. Kohler said that this could not be accounted for by slow trial-and-error stamping-in of a Thorndikian habit. He instead proposed *insight* as the underlying basis of sudden solutions. Kohler, who lived in Germany, was Estonian by birth, and protested the German treatment of Jews. He and some other German psychologists immigrated to the United States as World War II was approaching, where they found the behaviorist-dominated environment hostile to their mindset.

In the United Kingdom, the idea of the mind was kept alive and well through the work of psychologists studying memory, attention, and other cognitive processes. For example, in the 1930s Frederic Bartlett introduced the idea of organized knowledge structures called *schema* that people use to understand the world, an idea that is closely related to Gestalt principles of how the mind creates meaning. Schema also played an important role in Swiss psychologist Jean Piaget's theory of how children acquire knowledge.

Animal-behavior researchers in Europe tended to prefer the ethological approach over behaviorism. Ethology, in fact, emerged in part as a response to the lack of interest in evolution shown by behaviorists. Among the pioneers of ethology were Niko Tinbergen (Netherlands), Konrad Lorenz (Austria / Prussia), Karl von Fritch (Austria / Germany), and, in some accounts, Jakob von Uexküll (Estonia / Germany). Ethologists were more interested in behavior than in the mind, but they were less reluctant than behaviorists to

Schema

In psychology, the word *schema* describes a kind of mental structure, a set of *linked representations* (bundles of memories) that we use to conceptualize and understand our world and respond to it. Schema shape our perceptions, thoughts, and emotions, and guide the storage of new memories about them. We each have many kinds of schema related to categories of objects (animals, foods, tools, faces) and events (school, work, parties, funerals) that have recurred in life. You do not have to build your perceptions, thoughts, and feelings about dogs afresh every time you encounter one. Instead, schema are templates that you round out with details from the present moment. In a given situation, multiple schema are activated, and they change as the situation changes.

refer to internal factors (especially hypothetical states of the nervous system) when explaining behavior. After World War II, Tinbergen moved to the United Kingdom, and established ethology there.

Although some British animal researchers in the mid-twentieth century were taking a behavioristic approach to associative learning, by the 1970s cognitive elements were being included. This in part was due to renewed interest in the work of Kenneth Craik, a Cambridge colleague of Bartlett's, who, in the 1940s, expressed doubt about behaviorism. He argued that the control of complex behavior in humans and other animals depends on internal mental models of the external world. Craik writes:

> There are various theories—such as the stimulus-response theory, the theory of instincts and drives, and the theory of conditioned reflexes and modifiable responses. None of these seem to me to put the emphasis in the right place; the nature of the animal and human mind seems rather to be to copy its environment within itself in an active, dynamic, model. . . . If the organism carries a "small-scale model" of external reality and of its own possible actions within its head, it is able to try out various alternatives, conclude which is the best of them, react to future situations before they arise, utilize the knowledge of past events in dealing with the present and the future, and in every way to react in a much fuller, safer, and more competent manner to the emergencies which face it.

Even some behaviorists in the United States protested against the extreme views that dominated their field. For example, early on, Edward Tolman broke ranks with the stimulus-response approach. He noted:

> According to this "stimulus-response" school the rat in . . . the maze is helplessly responding to a succession of . . . external

> and internal stimuli (that) callout the walkings, runnings, turnings, retracings, smellings, rearings, and the like which appear.

Tolman proposed, instead, that instrumental behavior in animals is goal-directed, or purposive. Though an American, he was greatly influenced by German Gestalt psychologists, and like them, argued that complex behaviors could not be explained simply by chains of stimulus-response associations. Even more shocking to behaviorists' ears, Tolman proposed that cognitive factors inside the organism, including thoughts and expectations, were needed to understand purposive actions in animals and humans alike. In a classic study he found that rats that were allowed to explore a maze without receiving any reinforcement for doing so nevertheless later solved the maze more quickly. He proposed that the rats learned the layout of the maze during the first stage and then used this internal representation (a memory schema) when they navigated the maze in the second stage, where success led to food. He called this *latent learning.*

In later experiments Tolman found that rats can learn mazes using two kinds of strategies. Some acquire motor responses (stimulus-response learning), while others learn about the spatial layout of the environment. He coined the term *cognitive map* to account for how animals use spatial memory to navigate in their worlds. With a view conceptually quite like Craik's idea of a mental model, Tolman noted:

> The stimuli . . . are usually worked over and elaborated . . . into a tentative, cognitive-like map of the environment. And it is this tentative map, indicating routes and paths and environmental relationships, which finally determines what responses, if any, the animal will finally release.

Karl Lashley, who, like Tolman, had spent decades studying maze learning in rats, made a powerful and influential case against the stimulus-response focus of behaviorism at a symposium at Caltech in 1948. He argued that complex behaviors in animals or humans cannot be accounted for by mere chains of associations. Instead, they should be thought of as being hierarchically organized, rather than linear, and based on plans. He also pointed out that much of the underlying processing takes place unconsciously. This helped set the stage for a new science of mind, a cognitive psychology, that was not solely about consciousness. Lashley additionally noted that language must be factored into the explanation of human behavior, implying that there are limits to what can be learned from animals.

Meanwhile in Canada, Donald Hebb, a professor at McGill University, published in 1949 the influential book *The Organization of Behavior,* which broke with behaviorism. Hebb had trained with Lashley, and also believed that psychological processes like perception, memory, thinking, and volition could not be explained by stimulus-response habits. Further like Lashley, he called upon inner states of the brain as playing an organizational role in these psychological processes.

One of Hebb's students, Brenda Milner, was interested in the intellectual functions of the temporal lobe, and Hebb arranged for her to have access to patients who had undergone surgical removal of epileptic tissue from their temporal lobes by the McGill neurosurgeon Wilder Penfield. After completing her PhD, Milner continued to work on such patients at McGill. But her breakthrough discovery was on a patient known as HM, whose hippocampi were removed in a surgical procedure in Hartford, Connecticut, rather than in Montreal. After the surgery, HM's short-term memory lasted seconds to minutes but was not converted into new long-term memories of life's experiences. From this finding, the hippocampus came to be understood as crucially involved in forming and storing long-term memories. Milner's findings also highlighted the difference between what would come to be known as explicit and implicit forms of learning, with explicit forms being cognitively accessible conscious memories, and implicit forms referring to procedural memories that are learned and used non-consciously, like habits or skills.

Lashley's comment about language presaged a debate that erupted in the 1950s over language and that helped solidify cognition as the dominant theme of psychology. B. F. Skinner, in true behaviorist form, said that children acquire language by the differential reinforcement of *verbal behavior,* much like other animals learn their behaviors. The linguist Noam Chomsky, by contrast, argued that language emerges from an internal mental structure, a generative grammar that is unique to humans and that underlies all human mentation. Around the same time Jerome Bruner published a book called *The Study of Thinking.* He argued that concept formation occurs not by way of strings of associations, but instead as a mental process.

Also in the 1950s, around the same time as surgery on HM, the psycholinguist George Miller, who had leaned toward behaviorism early in his career, offered another significant challenge to stimulus-response psychology. This transpired in a famous paper on short-term memory titled "The Magic Number Seven, Plus or Minus Two: Some Limits on Our Capacity for Processing Information." In it, Miller showed that people can *hold in mind* (that is, in short-term memory) roughly seven items in the process of thinking and reasoning about the world. This was radical because behaviorists talked about learning, not memory. Miller went on to suggest that the fixed mental limit of short-term memory is easily circumvented by cognitive strategies such as *chunking,* in which information is bundled or grouped by recoding the information. For example, a typical phone number in the United States is ten digits long. It is easier to remember such a number by separating the ten items into three groups (area code, exchange, and number). Miller was also the lead author of a seminal 1960 book on cognition, *Plans and the Structure of Behavior* (the word "plans" in the title was code for cognition, but inclusion of "behavior" rather than "cognition" or "mind" would seem to reflect the lingering influence of behaviorism).

These various events in the 1940s, 1950s, and 1960s were occurring against the backdrop of an important development in the field of computer science. Researchers such as Craik, Alan Turning, Norbert Weiner, Claude Shannon, Herbert Simon, Allen Newell, and Marvin Minsky were making noises about similarities between the ways that machines and human minds process information.

What scientists call themselves often dictates how they approach their field. Watson and his followers, of course, called themselves behaviorists. And with the publication of Ulrich Neisser's 1967 book *Cognitive Psychology,* psychologists began to refer to themselves as *cognitive psychologists.* Although behaviorism persisted for a while, especially in animal research, by the 1980s even animal researchers were flirting with cognitive concepts. Today, the cognitive approach is dominant, though remarkably behaviorism has not been fully exorcised.

Revolution or Evolution

It is often said that the scientific study of cognition was the result of a revolution in the middle of the twentieth century. But George Mandler, a leading cognitive scientist in the early days, argued that there was no actual revolution, no insurgent force, that overthrew behaviorism. The demise of behaviorism, he asserted, was an inevitable consequence of Watson's founding manifesto that treated human and non-human animals as interchangeable in psychological research. The cognitive approach grew in influence over time, he said, because it gave psychologists interested in the human mind a viable alternative to behaviorism.

15

What Is Cognition?

A typical contemporary definition of cognition is "the mental action or process of acquiring knowledge and understanding through thought, experience, and the senses." Such a definition, though, begs the question of what *mental, thought,* and *experience* mean, because each is steeped in centuries of philosophical speculation.

For example, in Descartes's famous saying *cogito ergo sum, cogito* referred to *thinking,* which he equated with one's mental, or conscious, experiences. As a result, *mental, mind, thought,* and *experience* all came to be about human conscious states. Other animals were not part of the Cartesian conversation, because they were viewed as mindless *beast machines* (that is, mere neurobiological beings) that react automatically to stimuli.

Cognition and Consciousness

Cognition and consciousness are entwined. As a result, we cannot understand cognition independent of cognition. For this reason, consciousness in the general sense of being aware of what is on our mind will be mentioned here. But the deconstruction of my view of what consciousness is, and how it comes about, will be explained in the next and final part.

Descartes's view that everything mental was conscious was very influential in philosophy, even though some German philosophers, like Leibnitz, Kant, and Herbart, proposed the possibility of unconscious thoughts. And when psychology emerged in the late nineteenth century in Germany, consciousness was the main subject matter, but unconscious thought was again considered by some. Nevertheless, both animal and human psychologists coalesced around consciousness as a primary explanation of behavior. This made it easy for the behaviorists to make consciousness a one-size-fits-all source of disdain. Several decades later, cognitive psychologists displaced behaviorism, but not by making consciousness their main subject. They rose to the top by taking a smaller step away from behaviorism, emphasizing information processing rather than consciousness.

Specifically, early cognitive enthusiasts, inspired by similarities between computers and minds, focused on how information processing might create internal representations of the world that could be used in thinking and behavioral control. They acknowledged that conscious states could result from this cognitive processing, but they were more interested in the processing itself than in the subjective conscious experiences that sometimes follow from such processing.

Early cognitive psychologists tiptoed around not only consciousness but also unconscious states for fear of being seen as modern Freudians. They built on Karl Lashley's idea that our conscious thoughts are based on unconscious cognitive processes, but they took pains to note that the unconscious in cognitive psychology was not about the repression of troubling memories. It was about nonconscious processes that contribute to cognition.

In line with Lashley, cognitive psychologists in the 1960s and 1970s generally assumed that people lack introspective access to the cognitive processes that underlie conscious thoughts. For example, George Mandler noted that the cognitive processes that support conscious experience "are not available to conscious experience, be they feature analyzers, deep syntactic structures, affective appraisals, computational processes, language production systems, or action systems of many kinds." Similarly, George Miller said that thoughts "appear spontaneously in consciousness." To make his point, he used an anecdotal example: if you are asked your mother's maiden name, the answer simply appears in your conscious mind without any conscious understanding of how that happened.

In the 1980s, John Kihlstrom dubbed these unconscious cognitive processes the *cognitive unconscious.* For example, if you are asked to fill in the blank letter in the word *_urse,* you are much more likely to say *nurse* if you are looking at a picture of a hospital, but *purse* if looking at a picture of a bank. The picture primes relevant memories, and you infer the name of the word based on that. Such memories are in effect schema that help conceptualize situations.

While every conscious thought or decision is based on unconscious, or shall we say, pre-conscious, cognitive information processing, not all pre-conscious processing results in conscious states. Furthermore, not all unconscious processing is cognitive.

The acknowledgment by contemporary psychologists that humans have information-processing capacities that can take place both consciously and non-consciously, and that can depend on either cognitive or non-cognitive capacities, results in three general classes of behavioral control in humans that reflect three of our four realms of existence.

Three Classes of Behavioral Control in Humans

1. Not Cognitive and Not Conscious (neurobiological realm)

–non-conscious processes that control reflexes, innate behaviors, and habits

2. **Cognitive but Not Conscious (cognitive realm)**
   –non-conscious (including pre-conscious) cognitive processes that control goal-directed behavior
3. **Cognitive and Conscious (conscious realm)**
   –conscious mental states that control goal-directed behavior

Some might say that I have omitted an important category—Not Cognitive but Conscious—which includes, for example, the notion of *sentience,* a kind of primitive consciousness based on raw feelings. But this issue is best saved for Part V, where I will explore the conscious realm in depth.

Cognition and Working Memory

Working memory is central to our modern understanding of cognition. The term was first used by computer scientists in the 1950s to describe a component of a computer program that stores information temporarily for use in solving a problem. In psychology, working memory was initially used sporadically to describe short-term memory and its capacity to hold a limited amount of information in mind.

In the 1970s, Alan Baddeley transformed working memory into a prominent research topic, and gave the concept its current meaning. He succinctly defined working memory this way:

> The term working memory refers to a brain system that provides temporary storage and manipulation of the information necessary for such complex cognitive tasks as language comprehension, learning, and reasoning.

Baddeley divided working memory into a general-purpose component called the *central executive* and two specialized systems: an *articulatory loop* for language processing, and a *visuo-spatial scratchpad*. The central executive was said to be involved in the cognitive control and monitoring of information processing. It retrieved relevant long-term memories and orchestrated the maintenance of the selected information in a temporary, short-term memory state, allowing the information to be used in thinking, reasoning, planning, and deciding. While working memory and attention are very closely related, they are not quite the same—some aspects of attention do not require executive control by working memory.

Because working memory has a limited capacity, we can hold only a limited amount of information in mind at any given time. To create a more lasting form of memory, a long-term memory, requires rehearsing the information implicitly or explicitly, keeping it active long enough for that long-term memory to be stored. For example, if you are given a phone number to remember and you rehearse it sufficiently, you will be able to remember it later. But if someone asks you a question while you are rehearsing it, storage of the memory will be disrupted and you will likely forget the number.

Early theories emphasized covert verbal rehearsal as a maintenance mechanism. Because rehearsal mainly works for verbal information, another process called *attention-refreshing* is thought to be important because it can be used to maintain any kind of information. Although temporary storage of information was originally assumed to be a function of the central executive, these days various specialized systems (especially sensory and memory systems) are viewed as primary loci of temporary storage.

Joaquin Fuster, a leading researcher, pointed out the breadth of working memory content, explaining that it "can be sensory, motor, or mixed; it may consist of reactivated perceptual memory or the motor memory of the act to be performed, or both. It may consist of the representation of the cognitive or behavioral goal of the act."

An ongoing debate about working memory concerns whether it is modular (consists of domain-specific specialized processors and an executive controller) or unitary (consists only of domain-general processes). Attention plays a greater role in theories that assume domain-general processes than in those that emphasize modular ones, since it selects what is to be temporarily stored and maintained.

The amount of information one can temporarily hold in mind when doing mental work is referred to as *working memory capacity.* Research shows that individuals differ in this ability. For example, working memory capacity is correlated with an individual's score on general intelligence tests and other demanding tasks, and it degrades in conditions such as Alzheimer's disease and schizophrenia.

The information content of working memory has long been associated with conscious experience in humans. This has led to working memory being equated with consciousness. But it is now known that working memory involves both conscious and nonconscious processing.

In the rest of this book, I will use a broad version of working memory that treats it as a general mechanism of cognition. The essence of what I have in mind is summarized by Earl Miller and colleagues:

> Working memory (WM) is a key aspect of higher-order cognition. . . . It is a mental sketchpad for the short-term storage and top-down control of information. An essential feature of WM is flexibility. A wide variety of information, sensory inputs, decisions, recalled memories etc. can be selected, maintained, manipulated, and read out as needed.

Indeed, we use working memory when attending, perceiving, remembering, planning, making decisions, and behaving. With working memory we assess the situations in which we find ourselves. We also use working memory to construct our thoughts and feelings about ourselves, and to monitor and examine these. This ability to cognitively apprehend our own mental states is called *meta-cognition.* We also use our understanding of our own minds to infer the mental states of others, including other animals, a kind of meta-cognition called *theory of mind.*

Cognition and Memory

One of the most important functions of cognition, including working memory, is that it helps us to make sense of the world. And it cannot do this without long-term memory, and especially schema.

The contribution of long-term memory to working memory was nicely described by Steven Kosslyn, who said that working memory functions "require not only some form of temporary storage, but also an interplay between information that is stored temporarily and a larger body of stored knowledge." This body of stored knowledge is called semantic memory. We use working-memory executive functions to retrieve semantic memories as we try to understand situations and make decisions about them.

Semantic memories are about facts, such as what apples are, or where Italy is located in the world. By interacting as schema, they help us meaningfully conceptualize and categorize the objects and situations in which we find ourselves. Semantic memories also underlie the flexible use of language to express meanings non-literally, such as when we combine concepts to communicate metaphorically. Science, art, music, religion, politics, society, and culture are based on factual and conceptual semantic knowledge.

The Dual-System / Dual-Process Approach to Human Cognition

Contemporary cognitive theories often draw on the difference between intuitions and deliberations. Intuitions are fast, effortless, automatic, and unconscious, while deliberations are slower, more effortful, and conscious.

The distinction between intuition and deliberation was popularized in Daniel Kahneman's 2011 book *Thinking, Fast and Slow,* which proposes two systems of thought. According to Kahneman, System 1 includes innate capacities we've inherited from, and share with, other mammals. Kahneman notes, "We are born ready to perceive the world around us, recognize objects, orient attention, avoid losses and fear spiders." But his System 1 is not purely innate. It also learns associations. Kahneman's System 2, by contrast, allocates attention to activities that require mental work, and its functions are closely associated with agency, choice, concentration, and conscious experience.

A defining feature of System 2 deliberation is that it depends on working memory. Unlike intuition, it uses the deliberative powers of working memory to explore alternative decisions and actions. The downside is that deliberation is computationally and energetically expensive. System 1 is more efficient, but the energy involved in System 2 activities allows you to purposively or deliberatively pursue your goals.

Dual-system views have been around for a while. William James distinguished associative thinking from reasoning. More recently Jonathan Evans and colleagues have noted that some form of dual processing always accompanies conscious thought and behavior.

Evans and Keith Stanovich pointed out that because System 2 processes depend on working memory, they are a natural category. But System 1 processes have no obvious underlying psychological unifying factor. They are a grab-bag collection of processes, the only common features of which are that they do not belong in System 2 since they do not depend on working memory and are not conscious. Somewhat like Kahneman, but before him, Evans and Stanovich described features of System 1 processes as "implicit," "associative," "early evolved," "old brain," and "similar to animal cognition."

On the surface, this seems straightforward. But if we dig deeper, some issues emerge. For example, implicit, associative, early evolved, and old brain sound more like processes that belong to the neurobiological rather than the cognitive realm. Processes "similar to animal cognition" also seem misclassified, but in the other direction.

Specifically, because other mammals have working memory capacities and can think (recall Tolman), "processes similar to animal cognition" should go with other cognitive (working-memory-dependent) capacities. Recall that working memory has conscious and non-conscious facets. Consequently, just because something is fast and not conscious—like intuition—does not mean it is not cognitive; nor does it mean that it operates independent of working memory. Evans has, in fact, said that "most of the workings of System 2 are unconscious." I suggest that intuition is a fast, non-conscious form of mental processing. As such, it is cognitive and belongs with other cognitive processes in System 2, not in System 1 with reflexes and habits. The two-systems idea needs an overhaul.

Rethinking the Dual-Process Approach in Humans

I think we need a three-systems approach in which the various two-system processes are re-distributed. To do this, I build on the partition of behavioral control described earlier.

Turning Two Systems into Three

System 1: Non-Cognitive and Non-Conscious Behavioral Control *(neurobiological realm)*

- –reflexes
- –instincts
- –Pavlovian-conditioned responses
- –habits

System 2: Cognitive but Not Conscious Behavioral Control *(cognitive realm)*

- –non-conscious working memory
- –non-conscious deliberation
- –non-conscious inferential reasoning
- –non-conscious intuition

System 3: Cognitive and Conscious Behavioral Control *(conscious realm)*

–conscious working memory
–conscious deliberation
–conscious inferential reasoning

System 1 has two entry criteria. It consists of processes that are both non-conscious and non-cognitive. Included are neurobiological realm behavioral processes that control reflexes, instincts, conditioned responses, habits, and so on. As a result, all things cognitive are now in Systems 2 and 3. System 2 includes instances of behavioral control that are cognitive but not conscious, like intuition, but also non-conscious inferential reasoning and non-conscious deliberation, and non-conscious working memory. Each of these qualifies as components of Kihlstrom's cognitive unconscious, including non-conscious working memory. More generally, System 2 processes include any instance in which representations are constructed by using mental models. System 3, by contrast, is composed of the traditional System 2 processes that depend on conscious working memory, such as conscious deliberation and conscious inferential reasoning.

Some readers may be confused about how something can be cognitive but not conscious. Consider this example. In a typical conversation, there is some topic being discussed. Relevant memories bundled together as non-conscious active schema serve as templates of thought and speech. As a result, you can converse back forth without having to think consciously about exactly what you are saying (my System 2). But if the topic veers, or you disagree with the other person, then you probably need to consciously consider where to take things (my System 3), which results in a restructuring of the relevant schema, allowing you to slip back into non-conscious, cognitive thinking and talking (my System 2). This idea of non-conscious cognitive thinking and talking is a crucial part of my theory of consciousness, explained in Part V.

Dual Systems of Behavioral Control in Animals

By the 1970s, human psychology had essentially become cognitive psychology. But the lingering effects of behaviorism delayed animal researchers from joining the cognitive party. While new research was continuing to support the idea that cognition contributed to associative learning in non-human primates and rodents, these results were often quickly followed by studies showing that a stimulus-response explanation was just as good or better. Consequently, many animal psychologists remained committed to behaviorist principles and continued to reject the idea that some animals in some circumstances are cognitive creatures that pursue goals.

In the late 1970s Tolman's ideas were dusted off and put in the spotlight by John O'Keefe and Lynn Nadel in their book *The Hippocampus as a Cognitive Map.* It built on O'Keefe's observation that neurons in the hippocampus, a region of limbic allocortex, fired action potentials as rats explored a novel environment and formed memories about it. The firing patterns of the cells seemed to reflect the construction of an internal spatial map of the environment—in other words, the hippocampus was acting like the brain's GPS. O'Keefe shared the 2014 Nobel Prize in Physiology or Medicine for this pioneering work.

Even evidence for spatial maps in the brain did not convince hardcore behaviorists. The Cambridge psychologist Anthony Dickinson characterized the situation in terms of a methodological stumbling block, noting that it is extremely difficult to distinguish from the mere observation of behavior whether an animal is responding habitually or in a goal-directed way, since the behavior looks more or less the same for both. The challenge, then, was to design laboratory tests that could distinguish goal-directed behavior from stimulus-response habits.

One of these tests, by Dickinson and Bernard Balleine, involved a procedure called *reinforcer re-valuation.* In a typical study, rats were trained to acquire an instrumental behavior—like pressing a lever when a light is on—in order to receive a food reinforcement. Some days later, the food was devalued before the test began, either by overfeeding the rats that food, or by making them nauseous (by injecting a chemical, lithium chloride) after eating it. Rats that later (after the effects of satiation or nausea had subsided) continued to perform the light-signaled behavior despite the food having been devalued were said to have learned a habit (since the response was simply elicited by the stimulus and did not depend on the current value of the goal), whereas those that stopped performing the behavior were said to have learned a goal-directed response (since devaluing the food disrupted their motivation to work to obtain the goal).

In a sense, these studies were solidly associative, and hence consistent with the behaviorist roots of the researchers. That is, just as habits are learned when the reinforcer stamps in an association between the response and the reinforcing stimulus that elicits it (stimulus-response associative learning), goal-directed behavior is learned by forming an association between an action and its reinforcing consequences (action-outcome associative learning).

But Dickinson and Balleine offered a cognitive account of goal-directed learning in their rats. They argued that the key factor that drives goal-directed behavior in future situations is not simply the previously learned action-outcome association. Also important is the *memory*—the internal (cognitive) representation—of the value of the reinforcer to the animal. In conceptualizing things in terms of memory, Balleine and Dickinson broke with behaviorist teaching since memory, like cognition and consciousness, was a concept that behaviorists shunned because it was too internal, too mental.

Later studies by Balleine suggested that even in rats, the use of memory representations to evaluate the current value of the action-outcome association requires working memory. This paved the way for the habit versus goal-directed-behavior dichotomy to be seen as an animal version of the traditional dual-system view that was being applied to humans—that is, in animals and people alike, habits are traditional System 1 processes, while goal-directed actions depend on working memory, and hence belong to the traditional System 2. The question of whether rodents have conscious processes will be discussed in Part V.

16

Mental Models

Non-cognitive control is retrospective; it is based on inherited innate wiring or previously learned associations between stimuli. Cognitive control via working memory is prospective, or future-looking, allowing one to imagine novel solutions to problems without having to first test those via real-world actions. In other words, cognition is a set of brain processes that transcend mere trial-and-error learning by allowing the organism to mentally simulate the effectiveness of alternative actions without having to risk life or limb. These processes, grounded in working memory, construct mental models using internal representations based on long-term memory, including schema, to make predictions when perceiving, attending to, remembering, thinking about, deciding, feeling, and acting in the world, or when imagining what does not exist, and in trying to understand who we are. While simulations within mental models can, of course, be conscious, in this part of the book I will emphasize non-conscious mental-model processing and control activities that can, but that do not necessarily, lead to conscious content.

Mental Models and Reasoning

In 1943, in his book *The Nature of Explanation,* Kenneth Craik introduced mental models, an idea that would have important, but separate, influences on the fields of computer science and psychology. The next year, Craik was appointed the inaugural director of the Applied Psychology Unit at Cambridge University, a group established to use engineering and psychological expertise to develop technologies that would support the British effort in World War II. Just one year later, at the age of thirty-one, he died from injuries sustained in a bicycle accident.

Craik's line of succession would define major landmarks in the modern history of cognitive science. Upon Craik's death, his colleague at Cambridge, Frederic Bartlett, took over the unit. This was fitting since Craik's mental-model theory had been influenced by Bartlett's ideas about the role of schemas in thought and action. Then some three decades later, in 1973, Alan Baddeley became director of the Applied Psychology Unit, publishing in the same year his first paper on working memory. As we will see, working memory plays a key role in our understanding of mental modeling. Therefore, one cannot help but wonder whether the pioneering insights of Craik and Bartlett were in the back, if not the front, of Baddeley's mind as he went about the business of providing cognitive psychology with a new framework for understanding how cognition works.

Starting in 1983, the assistant director of the unit under Baddeley was Philip Johnson-Laird, who that same year published *Mental Models: Towards a Cognitive Science of Language, Inference, and Consciousness.* Johnson-Laird credited Craik for inspiration, and went on to spend much of his career developing a theory of how humans use mental models to draw inferences when reasoning. Not surprisingly, given Johnson-Laird's relation to Baddeley, he emphasized that mental models depend on working memory. For example, he noted that the mental-model theory assumes that the more information one needs to make an inference, the more difficult it should be to draw that inference. And because working memory, as we have seen, has a limited capacity, we often use only a subset of the knowledge available to reach conclusions. Consequently, inferences based on mental models can be very useful, but also faulty.

As an example of how mental models contribute to reasoning, consider the assertion "All artists are bakers." Johnson-Laird pointed out that this assertion implies the existence of a mental model representing the relation between the two sets of individuals. The holder of this model thus can infer that if someone is an artist, they are also a baker. But if the premises are, as in this case, untrue (all artists are not necessarily bakers), the inferences from such a model will lead to a false conclusion and result in errors of reasoning. Because the situations we reason about are not always black and white, Johnson-Laird's mental-model theory proposes that one can draw a conclusion of certainty, high probability, or mere possibility, depending on whether the conclusion holds in all, most, or some of the resulting models. If Johnson-Laird were asked today to comment on the societal implications of these simple principles, he might well construct a mental model in his own mind that infers that when social groups share inferences based on untrue mental models, the result is shared false beliefs (so-called alternative facts).

Johnson-Laird placed his theory of mental models into the System 1 versus System 2 context, stating that our System 1 intuitions make no use of working memory to hold intermediate conclusions in mind. Like other dual-system theories, then, his version suffers from putting intuition in the same category as reflexes and instincts. In my three-system approach described earlier, intuitions are part of the cognitive unconscious.

Mental Models and Goal-Directed Behavior

Given the dominance of dual-system approaches to cognition, Nathaniel Daw, Yael Niv, and Peter Dayan at University College London asked how the brain arbitrates between different control networks, especially networks underlying habits and goal-directed actions, when these disagree. For an answer, they turned to the field of machine learning. Researchers there sought to understand how different systems learn to choose actions that maximize rewards and minimize punishments by using errors to improve performance. Such algorithms are now part of everyday life: they underlie web search engines, face recognition by your smartphone, DNA sequencing, self-driving cars, computer-based diagnosis in medicine, and more.

Early machine-learning researchers proposed two kinds of learning. Model-free learning uses past data and accumulated associations to predict problem solutions. Model-based learning, by contrast, uses assumptions about the problem (a model) to make future predictions.

Daw and his colleagues brought the model-based versus model-free distinction into psychology and neuroscience, and used it to understand the differences between habits and goal-directed actions in humans. In the spirit of Dickinson and Balleine, they argued that model-free habit learning uses the association of an action with its past consequences to choose a response in the present, while model-based goal-directed learning assembles knowledge from memory to construct a mental model, and searches the content of the model in the process of deciding what to do.

Because model-based predictions are made on the fly, rather than being predetermined by the past, the decisions made allow flexibility when the value of goals or actions changes, or when a chosen behavior fails. In such situations, novel options can be explored within the model, and a new choice generated. In addition, a behavioral outcome learned in one setting (foraging for food) can be called on in other situations (when searching for water or escaping from harm). In using these models, working memory systems use executive control to retrieve and maintain those memories that generate options likely to be successful. In short, behaviors controlled by internal representations and mental models typify the cognitive realm.

Why does a brain equipped with mental models need habits? Because model-based processes require extensive information processing, they are computationally and energetically expensive. Model-free processes, by contrast, are less demanding. But they are rigid since they are controlled by the past. On the upside, habits are efficient when resources are plentiful. When they are scarce, the flexible choices made possible by mental models are a significant asset and worth the energy expenditure. While clearly different, model-free and model-based processes are not completely independent. For example, model-free associations contribute to model-based behavioral control the way reflexes contribute to intentional movements—they provide underlying support.

When it was first introduced to psychology and neuroscience, the model-based versus model-free idea was about the distinction between instrumental habits and goal-directed actions, with goal-directed actions helping to bring the notion of goal *value* into the reasoning processes mediated by working memory.

For example, across mammals the short-term maintenance of a goal's value, and the executive control of goal-directed action choices, came to be thought of as depending on working memory. Consistent with that idea, model-based decision-making correlates well with performance on working memory tasks, and stress impairs model-based working memory performance, shifting behavior toward habits. Further, individuals with greater working memory capacity demonstrate more resilience against the negative effects of stress, and people suffering from schizophrenia have impaired working memory and a reduced capacity for model-based decision-making. Finally, immaturity of model-based working memory systems is thought to explain why children persist in doing things that have stopped being beneficial to them, and why adolescents make impulsive, short-sighted choices.

Importantly, the distinction between model-based versus model-free behavioral control has emerged as a general way to characterize the difference between cognitive and non-cognitive processes. But there is a caveat. The machine-learning roots of this paradigm should not mislead us into thinking that these processes work in machines the same exact way they work in the human brain. For example, Daw and his colleagues started out using the computer metaphor to distinguish mental (cognitive realm) from non-mental (neurobiological realm) functions of the brain. But the more we learn about the mind, the less computer-like it seems. Indeed, given that at this point the human mind can do so much that machines cannot, increasingly it is being used as a metaphor to improve the information-processing capacities of machines by making them more humanlike. ChatGPT is a case in point.

Maps, Models, and a General Cognitive Framework

I had long assumed that Daw, Niv, and Dayan's model-based approach to cognition was inspired by Craik's mental-model musings. It would make sense, since they, like Craik, worked at the intersection between math and psychology in the United Kingdom, and Craik had made significant contributions to this kind of work. But when I contacted Daw to confirm my suspicions, I was surprised to learn that the group's effort to provide a cognitive interpretation of Dickinson's goal-directed behavior was more influenced by the tradition of Tolman and the work of John O'Keefe, a colleague of theirs at University College London who, as we've seen, in the 1970s had discovered evidence for Tolman-like cognitive maps in the hippocampus.

The model-based versus model-free distinction is a much-embraced approach for understanding cognitive control of behavior, one that includes instrumental goal-directed learning, decision-making, and working memory in humans and other mammals. It provides a compelling framework for understanding the mind and brain. Part of its strength and appeal, I believe, is its implicit fusion of the Craik and Tolman mental-model traditions, which have separately shaped the emergence of what we know now of as cognitive science.

Some of my colleagues, such as Paul Cisek, Luiz Pessoa, and György Buzsáki, have argued that cognition is simply a psychological category we've invented and that we should reduce it to a less abstract construct, or to a set of such constructs, that are more naturally tied to behavioral and brain evolution. I largely agree. And that is why I narrowly define cognition in terms of the capacity to construct models of the world and use these in thinking, planning, deciding, acting, and even feeling. In the next chapter, I explore the evolutionary roots of cognition, so defined.

17

Model-Based Cognition in Evolution

It is well established that mammals, when they pursue goals, can use internal representations to construct mental models that simulate the outcomes of possible future actions. The long, hard road to this conclusion has led many to assume that human cognition was the result of a defined thread of inheritance that started with early mammals. But some who study other animals have a different idea.

Are Non-Mammals Cognitive Creatures?

Over the past several decades, the mammal-centric view of human cognition, and the behavioral flexibility it allows, has been questioned. Specifically, accumulated research findings suggest that non-mammalian animals may also sport cognitive capacities. My main concern is with non-mammalian animals that are in our evolutionary past so that we can understand the origins of our own capacities. But first I will briefly consider research on animals that are not in our past.

Some scientists have claimed that protostome invertebrates, and even single-cell organisms, are cognitive beings, often basing those claims on a broad definition of cognition as information processing and learning. But given how I have defined cognition, simply having information processing and learning capacities is not enough for membership in the cognitive club. Nor is the ability to exhibit complex behavior.

Charles Abramson and Harrington Wells note that by using concepts like cognition, the "burgeoning field of invertebrate behavior is moving into what was the realm of human psychology." Complicating this effort, they say, is an inconsistent use of terms and a focus on similarities while ignoring the differences. The further removed a species is from us, they argue, the less applicable are ideas that humans have constructed about our own cognitive capacities.

Much of the work on protostome cognition is indeed guided by analogies with human, or at least mammalian, cognition. Because we humans understand what cognition is from our conscious understanding of our own mental capacities, analogy with human behavior is one of the only ways to approach the topic. For the most part, researchers doing this work are *not* trying to understand human cognition. Instead, they are interested in the capacities of their bee, fly, or octopus subjects, and they use analogies with humans to emphasize the complexity of the behaviors they study. That makes good sense, since the only connection that protostomes and vertebrates have is by way of the PDA living some 630 million years ago (see Part III). Whatever cognitive abilities present-day protostomes may have would have evolved independently in response to evolutionary pressures faced by their protostome ancestors.

Given this, it seems that the best way to empirically establish so-called *parallel evolution* (the separate evolution of similar capacities in different groups) of cognition in non-mammals would be to test them using experimental designs that have been used in mammals. As we've seen in mammals, habits and goal-directed actions are indistinguishable by mere observation of behavior, and are best separated empirically by using rigorous tests, especially involving reinforcer devaluation. If you want to know if protostomes exhibit goal-directed behavior, this re-valuation procedure would seem to be a crucial kind of test to use. But such procedures do not seem to have been used much in protostomes. Without employing such tests, researchers are left to infer humanlike cognition from factors such as behavioral complexity. Bees, for example, exhibit magnificently complex behaviors that seem consistent with the use of internal representations. But in the absence of devaluation-type studies, conclusions aligning with mammalian, including human, cognition would be inconclusive. That said, it's perfectly acceptable to propose the possibility of cognitive-like processes and internal representations in the absence of such tests, so long as the conclusions are acknowledged to be hypotheses.

Although reptiles are not in the direct evolutionary past of mammals, they are potentially relevant to humans because they share an amniote ancestor with early mammals via extinct therapsids. Amphibians and fish are unquestionably relevant since they are in the direct past of mammals. If fish and amphibians have cognitive capacities, it would challenge the idea that mammals invented the cognitive traits, and instead suggest that mammals simply modified capacities inherited from their more ancient vertebrate ancestors.

A number of claims have been made about cognition in nonmammalian vertebrates. While issues have been raised about the findings, gaps in knowledge prevent firm conclusions. Under ideal conditions, comparisons should be made using the same kinds of tasks as those used in studies of mammals, with modifications, of course, tailored to the species and its sensory-motor endowment. So far, there is a paucity of research using classic devaluation tests, but other related approaches have been used. Overall, the evidence supports instrumental habit learning, but not model-based instrumental goal-directed action learning in fish, amphibians, and reptiles.

Spatial mapping has been studied far more extensively in non-mammalian vertebrates than has goal-directed behavior using devaluation tests. And the results have provided considerable evidence that fish, amphibians, and reptiles use spatial maps in navigating the world and foraging in it. This would seem like slam-dunk evidence for cognition in these animals. But is it?

In his book, *A Brief History of Intelligence,* Max Bennett summarizes the state of the field. He starts by noting that "an agent's current state can be defined as a location in space, and the actions it associates rewards with can be defined by the next target locations, which thereby would generate a homing vector from the current location to the next target location." He points out that this kind of behavior has various adaptive properties, such as being insensitive to small changes in starting locations and / or paths. It does not, however, involve a "playing out" or simulation of state transitions. Bennett's bottom-line point is that spatial information can be used either in a model-free or model-based way. Unless an animal uses a mental simulation (a mental model) to evaluate the consequences of moving from one point in the spatial map to another, it is not engaging in a cognitive, model-based solution. This applies to all animals. And as Abramson and Wells noted earlier, the further from humans an animal is evolutionarily, the more difficult it is to use analogy with human behavior to draw conclusions.

From the evidence he reviewed, Bennett concluded that mammals have the ability to engage in model-based spatial learning, whereas lower vertebrates and protostome invertebrates are likely limited to model-free reinforcement learning and behavioral control. He nevertheless awarded lower vertebrates with other novel reinforcement-learning features that distinguish their model-free capacities from the model-free capacities of protostomes. One is the ability to learn about not only the presence of stimuli, but also their absence (so-called omission learning), and the other is the ability to learn about the timing of reinforcers. The limited extent to which these processes have been use in studies protostomes restricts the conclusions that can be drawn.

What about birds? It has long been known, especially from anecdotal evidence, but also from experiments, that some birds, especially parrots and corvids (part of the crow family), have impressive abilities. Writing about bird cognition, Nicola Clayton and Anthony Dickinson have argued that corvids use internal representations of where and when they stored food, and can make choices based on freshness—that is, they retrieve and consume older items before newer ones. Much research, including devaluation studies, suggests that cognition in corvids rivals that in primates. But the fact that such capacities are present only in some kinds of birds, and seem to be lacking in the reptile ancestors of birds, indicates that birds and mammals probably evolved these similar capacities separately, undoubtedly because of distinct selective pressures in their past.

In sum, mammals and birds pass cognitive tests of goal-directed behavior with flying colors, but lower vertebrates have trouble. None of this completely rules out the possibility that some kind of model-based cognition is present in fish, amphibians, and reptiles, or even in protostome invertebrates, especially because there is a built-in bias that results from using mammalian standards to assess non-mammals.

Some will say that cognition can be inferred in reptiles, fish, and / or protostomes from instances of complex behavior. In response I would say that defining cognition in terms of complex behavior is inadequate. Cognition is about mental processes. Complex behavior can be, but is not necessarily, based on mental processes. As we've seen, just because a behavior looks like it is goal-directed does not mean it is based on internal representations, and hence cognition.

Robert Barton, a leading evolutionary neuroscientist, recently made the point succinctly and pointedly in relation to research on the minds of octopus. He noted that the balance of emphasis seems wrong in the field. Rather than trying to find human-like mental states in octopus, researchers should be trying to understand the sorts of cognitive processes that are likely mediated by the radically different, independently evolved nervous system that comes with the particular kind of body of the octopus and its phylogenetic history.

Much empirical and conceptual work is currently under way in non-mammalian vertebrates and protostomes. It is therefore possible that compelling evidence for some form of model-based cognition will be forthcoming in these animals. But as one tantalizing fact suggests, maybe not.

Why Mammals and Birds?

Mammals and birds both inherited model-free precursors of model-based cognition from their respective vertebrate ancestors. But they seem to have separately invented model-based cognition on their own by modifying model-free mechanisms. And the reason they, and only they, have robust cognitive capacities may be a consequence of a specific set of body plan features that they both possess.

As discussed in Part III, the therapsid ancestors of mammals evolved a new feature—legs positioned directly underneath their trunks that could move parallel to their vertebral column. As a result, they could breathe and run at the same time when escaping predators or capturing prey. Because this behavioral adaptation required more energy, therapsids, including mammals, had to consume more food, but also had to take in more oxygen to make the extra energy, compared to almost all other groups of animals. Birds are the exception. They also evolved legs under their trunks, and reaped the benefits to foraging and defense.

The heat underlying these higher metabolic rates meant that mammals and birds were warm-blooded; that is, capable of maintaining their internal body temperature through metabolism. And warm bloodedness, or endothermy, may be the key to understanding the evolution of model-based cognition in mammals and birds. Specifically, model-based cognition may have evolved as a way to obtain the fuel required for their high metabolic rates.

But endothermy does not guarantee model-based cognition, since all birds are endothermic but only some have model-based cognition. A few non-mammalian vertebrates, and even a few protostome invertebrates, are endothermic in a situational sense but do not have the continuous full-body metabolic endothermy that mammals and birds have.

In an adult human, the brain accounts for only about 2 percent of total body weight, but 20 percent of oxygen use at rest. And when engaging in energy-demanding behaviors, oxygen use goes up considerably in the body and brain to support the metabolic demands of the behavior. But higher metabolism requires more fuel in the form of food, which, in turn, requires that more time be spent on foraging. And foraging is not just a source of fuel, it is also a major fuel-burning activity.

Advantages might therefore accrue to organisms that could use model-based abilities to make plans about when (time of year and time of day) and where (locally and distally) to forage for what (perishable or durable food) in relation to predictions about energy expenditure and energy prospects in the near- and long-term. Indeed, Elizabeth Murray and her colleagues proposed that endothermy evolved to support the energy required to be able to flexibly plan and make decisions. These abilities are especially useful in times of food scarcity, which brings with it greater competition for resources and increased risk. In bountiful times, warm-blooded animals, like cold-blooded ones, can routinely rely on neurobiological realm behaviors, such as area-restricted search and learned foraging habits, driven by Pavlovian-conditioned incentive stimuli. But times of plenty can also have challenges. When situations arise, such as a sudden change of weather, or an unexpected encounter with a predator, warm-blooded animals can rapidly switch to a model-based strategy. Cold-blooded animals, lacking this luxury, it seems, are stuck with their neurobiological realm options. Even if it turns out that they have talents that can be construed as cognitive-like, or even truly cognitive, they would not have the metabolic means to generate the energy mechanisms required to support model-based flexible planning and decision-making that allow true endotherms to acquire the fuel needed to maintain internal body temperature in the precise range necessary for metabolic homeostasis 24/7.

If indeed goal-directed planning and decision-making using internal representation and mental models arose symbiotically with endothermy, as this discussion implies, it would help explain why evidence of robust cognition is difficult to obtain, and controversial, in cold-blooded animals. Even if it turns out that they have some cognitive talents, they would not likely have the energy resources needed to support the kind of model-based flexible planning and decision-making possible in warm-blooded animals.

Models and Only Models

We've arrived at a mental-model view of mammalian cognition and have concluded that mammals and some birds each invented this capacity on their own, rather than having inherited it from their lower vertebrate ancestors. But, as I said earlier, they did not invent it out of whole cloth. Instead, under selective pressures related to the foraging requirements of being warm-blooded, they invented model-based cognition by modifying mechanisms that they inherited from their ancestors, especially the mechanisms of associative learning and habit formation.

The model-based versus model-free approach, when properly applied, captures what it means to be a cognitive creature. For this reason, model-based cognition will be the anchor for my exploration, in the next chapter, of the evolution of cognition in mammals.

18

Foraging in the Mind

Mental modeling is a widespread feature of mammals today. It is not known when it emerged: was it a defining feature when therapsids diverged from amniotes or when mammals diverged from early therapsids, or did it come about when mammals diversified after the demise of the dinosaurs?

Regardless of when cognition first appeared, it continued to develop as mammals faced additional challenges and evolved different body plans, as when primates diverged from ancestral mammals, apes from ancestral primates, and humans from other great apes. For this reason, I will separately discuss mental modeling in non-primate mammals (focusing on rats), non-human primates (focusing on monkeys), and great apes (focusing on humans). In so doing, I will treat each of these three groups as if it possesses a singular capacity for mental modeling that differs from the capacities of their ancestral groups. This is clearly a simplification, but one justified by the fact that cognition is more similar within each group than between them.

The Importance of Foraging

Considerable evidence points to a connection between selective pressures related to foraging and the cognitive capacities that various mammals evolved. Adaptations that help organisms acquire food more efficiently are highly favored by natural selection. And each kind of mammal developed different cognitive abilities as their body plans changed in response to pressures in their niche.

Let's start with the first mammals to diverge from amniote therapsid ancestors about 210 million years ago. These creatures foraged within small areas close to their home base. This was due to their limited visual capacities, nocturnal lifestyle, and dependence on olfactory cues that decay over short distances. In novel territories, they relied on specific sensory cues (innate or conditioned stimuli) for identifying and evaluating food sources and assessing risks. In other words, like their lower vertebrate ancestors, they used model-free behaviors in such situations. Also like their vertebrate ancestors, with repeated visits to an area, they could store memories of the spatial layout and landmarks that signify food or danger. But in contrast to lower vertebrates, they were able to use a mental model of the territory and its landmarks to store and retrieve memories of past payoffs. This allowed them to evaluate and make predictions about the probability, risk, and energy cost of procuring food in relation to their current needs and expectations about the future.

About 55 million years ago, primates diverged from the small insectivore mammals that had survived the climatic change that killed off the dinosaurs. The first primates, prosimians (examples today are bush babies and lemurs), made their living in the tree branches of the rain forest, where they were protected from carnivorous, ground-foraging mammals. They had forward-facing eyes, which gave them binocular vision and the ability to see objects in perspective and detail. Both their hands and their feet could grasp. Because their movement was propelled mainly by their legs, their hands were free to manipulate items as they searched for the nuts, fruits, and other kinds of vegetation that made up their diet. Because of their exquisite hand control, prosimians were described by Todd Preuss as *manual foragers.*

Monkeys diverged from prosimians about 35 million years ago, retaining prosimians' eye-hand coordination and adding new skills needed to survive under changing climatic conditions. Specifically, the food required to supply their high energy needs in the forests was becoming scarce due to global cooling. Slowly, their vision improved. They could discern color and detail and recognize objects at a distance. As these capacities evolved, they were able to shift to a diurnal (daytime) lifestyle, engaging in terrestrial foraging over large areas that covered thousands of hectares (for reference, four thousand hectares is about twelve square miles). Just as prosimians were manual foragers, monkeys were *visual foragers.*

Under the pressures of wide-range foraging, monkeys acquired the ability to remember the layout of complex expansive foraging territories and the visual features of landmarks that signified locations where useful and harmful circumstances were encountered. To effectively and efficiently forage, they had to be cognizant of the season, plan what food to focus on each trip, consider where it was likely to be found, and decide what time of day might be most productive for procuring it. Such planning abilities helped minimize errors in food collection and risk assessment, allowing monkeys to better adapt to limited and unpredictable food supplies and the presence of carnivorous terrestrial predators. Underlying these mental modeling abilities was the capacity for object recognition (semantic memory) in relation to spatial maps, and the capacity to select and maintain sensory semantic representations and schema in working memory when making decisions about which goals to pursue and how to achieve them.

Some 25 million years ago, apes appeared and diverged into two major groups. Lesser apes (gibbons) were similar in size and lifestyle to their monkey ancestors. Present-day ones reach at most seventeen or so pounds. The great apes (bonobos, orangutans, gorillas, chimpanzees, and humans) diverged about 18 million years ago. They were considerably heavier—for example, bonobos today weigh as much as 110 pounds, and gorillas can be more than five hundred pounds. Because of their large bodies, they had to use all four limbs to move on land. This resulted in even higher energy needs than monkeys.

Apes inherited from monkeys the ability to form mental models based on specialized representations of objects in visual space, but they added their own cognitive inventions. For example, Elizabeth Murray and colleagues proposed that great apes evolved a novel general cognitive capacity that added elaborated forms of conceptualization to semantic memory and top-down executive control. By accessing stored knowledge in this way, great apes could engage in inferential reasoning across categories. They concluded that a kind of general "intelligence" arose through this ability to extrapolate from specific scenarios (objects in a spatial layout) to novel situations. Recent studies in fact suggest that, like humans, chimpanzees are able to delay gratification, an ability that correlates strongly with performance on tests of general intelligence in humans.

Mental Search

Humans began to diverge from other great apes around eight or so million years ago, and began to exist as several unique species about five or six million years ago. They lived in East Africa and foraged in the rain forest and the savanna. Between 3.5 and 2.5 million years ago, climatic changes led to movement of the continents, creating the Mediterranean Sea as well as the high mountains and deep lakes, especially in East Africa. This *rift system* formed a barrier that prevented access to the savanna and its fruits, and those living there became extinct, but those in other areas, like East Africa, were unaffected. This would come to be known as the *cradle of civilization*.

With their upright posture, speed on their feet, ability to swim, and reduced body hair, East African humans were able to forage both in the rainforests and in the hot, dry conditions of the savanna, where there were abundant sources of nuts, but also meat from ungulates. Killing large animals and fighting off competitive carnivores was aided by the invention of stone tools, such as axes, around three million years ago. By 1.5 million years ago, fire was controlled and food, especially meat, began to be cooked.

In emphasizing the role of foraging in the evolution of cognition, I have focused on how ecological factors, such as scarcity and risk in finding food, shaped the development of mammalian cognition. Another set of important factors that influenced cognitive evolution in humans was social in nature, and included cooperation, communication and language, theory of mind (the ability to make assumptions about the minds of others based on knowledge about how one's own mind works), and cultural transmission of knowledge.

For example, having a theory of mind may have allowed hunters separated in space to create mental models in which they could anticipate what each was likely to do as the hunt progressed, and verbal communication would have allowed the day's successes and failures to be shared and stored as memories that could be incorporated into the mental model of the hunt and enhance the probability of success on the next excursion. By the same token, foraging enhancements, such as tool use, likely led to efficiencies in finding food, which may have made more time available for social sharing of information and social bonding in less transactional ways.

This is not to say that social factors had no effect on other primates. Primates are in general highly social creatures and social pressures have also shaped their cognitive capacities. As Amanda Seed and Michael Tomasello note, there are important differences between monkeys and humans in terms of complex cognitive skills. But more basic capacities for cognitive processing and mental representation, which are used when interacting with the physical and social worlds, are also possessed by other primates.

An obvious difference between humans and other primates is language. The philosopher Peter Godfrey-Smith argues that complicated mental processes occur in animals without speech. This is obviously correct, but not the point. Daniel Dennett nailed it by saying that while language is not necessary for cognition, cognition without language is not the same as cognition with it. Language, Dennett said, creates tracks on which thoughts can travel. Birds make sounds to warn of danger or attract mates. Monkeys and apes have sounds for different predators (cats versus hawks, for example). But only humans flexibly indicate things about the present, past, and / or future by using syntax to convey to other humans exactly *when* and *where* a particular kind of predator was seen today, and then to discuss the implications with others so as to make a plan for how to act in the future.

Language is the ultimate multimodal system—it allows you to know that an apple on your kitchen table, a picture of an apple in a magazine, the letters A, P, P, L, and E in print or hand script, the sound of the spoken word *apple,* and the taste or smell of an apple all refer to the same thing. In the brain, language involves a sophisticated multimodal convergence system spread across the temporal, parietal, and frontal lobes.

An important conceptual capacity enabled by language is hierarchical relational reasoning. This is the ability to draw inferences about similarities and differences between superficially distinct things—for example, a rabbit hole and a hotel. Derek Penn, Keith Holyoak, and Daniel Povinelli put it this way:

> Although there is a profound similarity between human and nonhuman animals' abilities to learn about and act on the perceptual relations between events, properties, and objects in the world, only humans appear capable of reinterpreting the higher-order relation between these perceptual relations in a

> structurally systematic and inferentially productive fashion. In particular, only humans form general categories based on structural rather than perceptual criteria, find analogies between perceptually disparate relations, draw inferences based on the hierarchical or logical relation between relations, cognize the abstract functional role played by constituents in a relation as distinct from the constituents' perceptual characteristics, or postulate relations involving unobservable causes such as mental states.

Another language-related capacity is the ability to think recursively—to think in terms of interrelated sequences. According to Michael Corballis, recursive reasoning is "a ubiquitous property of the human mind and possibly the principal characteristic that distinguishes our species from all other creatures on the planet." Recursive reasoning is required to understand a sentence like "The dog that chased the cat that killed the rat." Humans can handle about five levels of recursion. Thomas Suddendorf offers this example: "A prisoner serving a life sentence for a crime committed in the heat of the moment, for instance, may think to himself: 'for the rest of my life (level 5), I will look back on my past (level 4) and regret the fact that I failed to anticipate (level 3) that I would regret my past decision (level 2) to commit that crime instead of not committing the crime (level 1).'" Young children and non-human primates seem to manage only a single level. Corballis explains that recursion is also necessary for imagining what others are thinking. He argues that other animals can perceive, think, know, and feel, but do not have the ability to recursively perceive, think, know, and feel what others are perceiving, thinking, knowing, and feeling. Michael Tomasello's research supports the idea that even chimpanzees do not have a full-blown humanlike theory of mind. Dwight Read obtained evidence for why monkeys and great apes lack complex recursion—they do not have the working-memory capacity required to hold multiple concepts in mind simultaneously. While this is certainly partly correct, language, particularly syntax, is also crucial for our ability to think across multiple levels.

A capacity that decidedly differs in humans and other animals is the ability to imagine possible future outcomes in one's own life, a capacity variably called mental time travel, prospection, self-projection, and future simulation. Suddendorf and colleagues take pains to distinguish this kind of cognitive prediction process from the more basic prediction processes that underlie associative learning, such as reinforcement learning based on the detection of prediction errors. The ability to mentally time travel to your future has not been demonstrated in any other animal, and as we will see, it underlies what is believed to be a unique form of consciousness in humans called *autonoesis.*

Thomas Hills has used the term *cognitive foraging* to refer to the ability to search flexibly through the mind for solutions to novel problems. If Hills is correct, cognitive foraging evolved by modifying the mechanisms of good old-fashioned foraging for food. The expression cognitive foraging is, in one sense, a metaphorical description of how we think. But in another sense, it captures the essence of how we use working memory and its executive functions in cognitively controlling goal-directed behavior.

Humans are, at least so far in the evolutionary landscape, the ultimate cognitive foragers. As when foraging for food, in mental foraging we have to balance exploitation of what exists now versus what might be found by further exploration by assessing the value (costs and benefits) of alternative opportunities based on internal goals and plans. Monkeys have a version of this capacity that allows them to mentally model and weigh current versus alternative goals, and apes add sophisticated conceptual abilities to that, but only humans can monitor the significance of multiple goals in parallel and switch among them. Humans inherited cognitive abilities from ape ancestors. But we have supercharged these capacities by adding language, symbolic thought, hierarchical reasoning, recursion, theory of mind, novel social skills, complex culture, art, music, religion, science, and medicine.

19

The Cognitive Brain

Brain structure and function tend to co-evolve in symbiotic steps. And sometimes small changes can have big consequences, leading to a major transition in behavioral control. One of these transitions, as we saw, occurred when bilaterals in the early Cambrian era invented model-free instrumental habit learning. They did this by combining dopamine-powered mechanisms of area-restricted foraging with associative learning mechanisms. Later in the Cambrian, invertebrate chordates inherited this habit mechanism and passed it on to vertebrates, where it was wired into the dorsal striatal region of the basal ganglia. Another major behavioral transition, and the one we were concerned with in Chapter 17, occurred when mammals co-opted model-free habit mechanisms in order to invent model-based goal-directed behavior to support endothermy. Here, I turn to the brain mechanisms of model-based cognition.

The Forebrains of Early Mammals

The brains of early vertebrates, as we've seen, possessed precursors (homologs) of the major components of the mammalian forebrain. These included areas involved in sensory processing (sensory neocortex), motor control (motor neocortex), navigation based on spatial memory (hippocampus), encoding of the value of innate or Pavlovian-conditioned stimuli encountered while foraging (amygdala and ventral striatum), acquiring instrumental habits (dorsal striatum), and coordinating visceral control with behavior (hypothalamus).

As mammals were emerging from therapsid ancestors, the greatest changes in the forebrain involved the cerebral cortex. According to Jon Kaas, arguably the leading expert today on the evolution of the mammalian cortex, the forebrain of non-mammalian vertebrates was transformed from a thin pallial layer into a thick, multilayered cerebral cortex with many specific functions (the pallium of ancestral vertebrates is the precursor of the mammalian cortex). The changes in mammals are accounted for by increases in the number of neurons in individual areas. With more neurons, the areas increased in size, were differentiated into multiple subareas, became more interconnected (both internally and with other areas), and took on new functions.

Kaas has pointed out that a number of the cortical areas present in early mammals were retained by all later mammals, including the sensory and motor neocortex, medial meso-cortical (limbic) prefrontal areas (orbital, medial, anterior cingulate, and insula), temporal lobe meso-cortical areas (rhinal cortices such as the perirhinal and entorhinal cortex), and allocortex (hippocampus and olfactory cortex).

Although the most impressive changes in early mammals compared to their amniote ancestors involved the cortex, subcortical forebrain areas also expanded. For example, the striatum of premammalian vertebrates consisted of just two main regions: the dorsal and ventral striata. In mammals, these striatal regions increased in size and expanded into distinct functional subareas. Other subcortical areas, such as the amygdala, also expanded in size and differentiated into multiple subareas.

All cortical and subcortical areas have subregions that typically contribute to different functions. When these subregions are taken into account, the number of functional areas that have long existed in the mammalian forebrain increases significantly.

Brain Mechanisms of Model-Based Behavioral Control in Non-Primate Mammals

As Tolman observed and many later confirmed, rodents can switch between using habits or goal-directed behaviors, depending on the momentary demands of the situation. One of the key forebrain changes in mammals may, therefore, have been the splitting of the dorsal striatum to include separate lateral and medial parts. The lateral part retained the classic striatal role in instrumental habit learning, and the medial part came to be a central component of the goal-directed behavior circuitry. This segregation is what allows mammals to behave rigidly, with habits, in response to stimuli learned about in the past, or flexibly in a goal-directed fashion in novel situations. Their ability to choose whether to behave habitually or flexibly was a tremendous advantage for mammals trying to keep their warm-blooded bodies well fed and regulated, whatever the environmental conditions.

Like the habit circuit centered on the lateral part of the dorsal striatum, the goal-directed circuit centered on the medial part is reciprocally connected with sensory and motor areas of the neocortex and receives inputs from midbrain dopamine neurons. But the goal-directed circuitry is considerably more complex (Figure 19.1). For one thing, in addition to connections with the sensory-motor cortex, it is interconnected with the prelimbic region of medial prefrontal cortex. It also requires an additional basal ganglia circuit, one centered on the ventral striatum (also known as the nucleus accumbens), and involving the orbital, ventromedial, and insula PFC, as well as the amygdala.

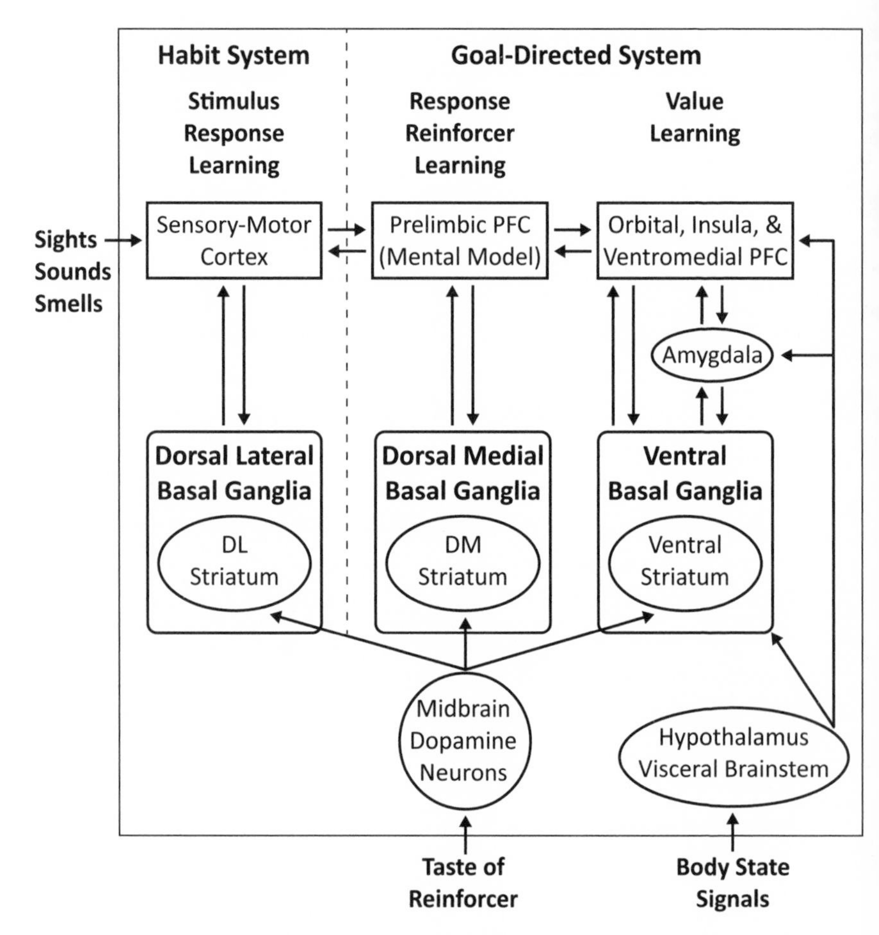

Figure 19.1. Goal-directed versus habit systems in mammals

The interconnected circuits in this collection of forebrain areas account for the main features of goal-directed behavior in non-primate mammals: (1) the formation of an instrumental response-reinforcer association (sensory-motor cortex, the medial part of dorsal lateral striatum, and the midbrain dopamine neurons), (2) the use of Pavlovian conditioning to incentivize the value of stimuli that predict reinforcers (amygdala and nucleus accumbens), (3) the retrieval of those incentive values from long-term memory (orbital PFC and amygdala), (4) the assessment of the present value of the reinforcer in relation to what the body needs (the orbital PFC, insula PFC, ventromedial PFC, amygdala), and (5) the use of working memory to temporarily store information needed to assess various values and actions and make choices (the prelimbic PFC). The combined effect of the various interactions among these processes is the flexible control of action.

Research on goal-directed behavior in non-primate mammals, for the most part, has involved laboratory studies, often in small experimental training boxes that lack ecological relevance to the animals. It is interesting that the hippocampus does not figure prominently in this literature, perhaps because navigation and spatial memory are not key factors in such limited environments. By contrast, in more complex situations in the lab, such as learning spatial mazes, or in the wild, where goal-seeking foraging occurs in open spaces, the hippocampus plays two important roles.

First, as we saw earlier, foraging can be model-free (when using area-restricted search and / or learned habits) or model-based (when using internal representations of the world to guide goal-directed behavior). Both strategies use spatial maps in navigation. Neurons in the hippocampal formation called *place cells* and *grid cells* represent the momentary location of the animal in space. Using neural signals from these cell groups these complex ensembles of hippocampal neurons generate possible spatial paths toward a goal. Foraging is model-based when, and only when, these spatial maps are part of a mental simulation of the possible outcomes.

Second, spatial maps are made and maintained by interactions between the processing of visual stimuli by circuits in the visual cortex with visual spatial processing circuits in the parietal lobe, and both of these with hippocampal spatial memory circuits. Long ago, Brenda Milner, who first implicated the hippocampus in memory, noted that spatial memory is a primordial example of a more general hippocampal capacity for remembering relationships between objects and the events in which they are embedded.

Indeed, when foraging, animals must use object memory to remember what objects are, where they are located, and how they relate to other objects and to reinforcers. Object memory—which depends on the visual cortex, perirhinal cortex, and the hippocampus—is believed to have evolved in conjunction with spatial navigation capacities.

Mental model simulation of possible outcomes using spatial maps and goal values is achieved by interactions between object memory, stimulus and response valuation, spatial mapping, and working memory circuits. Together, these make it possible for mammals to hold stimulus and response values in working memory while choosing how best to act, including where to go, when to go, how to get there, and what to do when there.

The Prefrontal Revolution

The basic forebrain areas and circuits underlying goal-directed behavior in non-human primates are essentially the same as those described earlier in rodents. But model-based cognition went to another level in primates as they diverged evolutionarily from their mammalian ancestors. For one thing, areas of the neocortex further expanded in size and complexity. While early mammals had about twenty functional neocortical areas, early primates had forty to fifty functional areas. As was the case in early mammals, the cortical expansion in early primates was due largely to an increase in the number of neurons in forebrain areas, which fostered corresponding increases in interconnectivity between neurons within areas and between neurons in different areas.

Particularly noteworthy was the expansion of prefrontal neocortex in primates. In lower mammals, you'll recall, the PFC primarily consists of meso-cortical areas, including the anterior cingulate, prelimbic, ventromedial, and orbital areas. In primates, these meso-cortical areas exist in expanded forms that include the differentiation of areas into subareas and increased connectivity with other areas. As in lower mammals, meso-cortical PFC areas in primates are mostly located in the medial walls of the two hemispheres.

Most important, primates evolved a large set of novel neocortical PFC areas. Because of a distinctive cellular feature in one cell layer, these areas are collectively called the *granular prefrontal cortex,* a designation first used in 1909 by the German anatomist Korbinian Brodmann. Brodmann concluded, and many contemporary authorities (including Todd Preuss, Steven Wise, and Richard Passingham) agree, that granular PFC is a novel add-on in primates, a primate invention that has no exact counterpart, no homolog, in non-primate mammals.

Granular PFC consists of several distinct areas that are located mostly on the lateral surface of the PFC, but two extend over to the medial side, where they abut meso-cortical areas that lack strong granulation. I will refer to the latter medial areas as *sub-granular PFC areas.* The key areas of granular and sub-granular PFC in humans are listed here and are depicted in the human brain in Figure 19.2.

Key Granular and Sub-Granular Areas of PFC of Humans

Granular (Neocortical) PFC	**Sub-granular (Meso-cortical / Allocortical) PFC**
Anterior pre-motor cortex	Anterior cingulate
Dorsolateral / ventrolateral	Prelimbic
Frontal pole (lateral and medial)	Ventromedial
Lateral orbital	Medial orbital
Insula	Insula*
Dorsomedial	

*sub-granular insula is the only allocortical area in this list

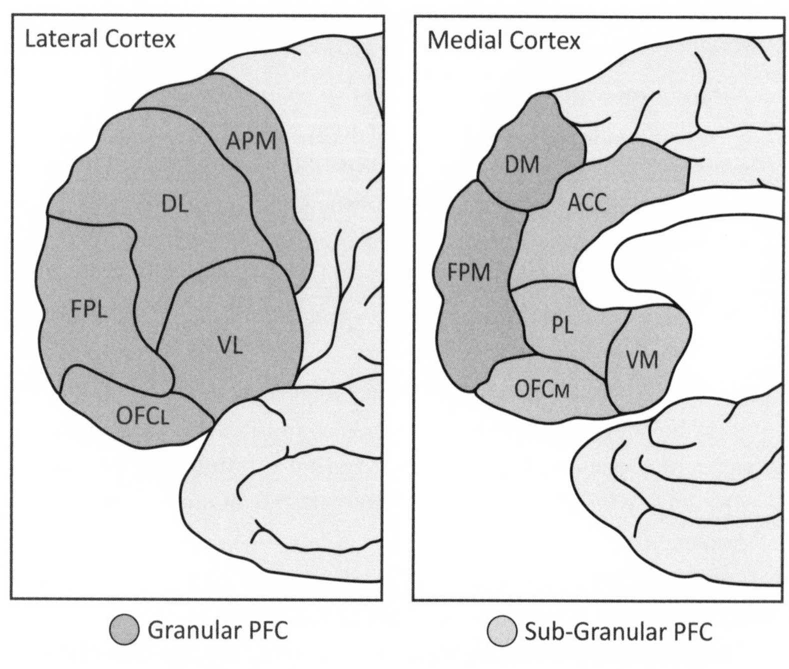

Figure 19.2. Granular and sub-granular prefrontal cortex in the human brain. The lateral prefrontal cortex includes the anterior premotor cortex (APM), dorsolateral prefrontal cortex (DL), ventrolateral prefrontal cortex (VL), lateral frontal pole (FPL), and lateral orbito-frontal cortex (OFCL). The medial prefrontal cortex is composed of the dorsomedial prefrontal cortex (DM), anterior cingulate prefrontal cortex (ACC), medial frontal pole (FPM), prelimbic cortex (PL), medial orbito-frontal cortex (OFCm), and ventromedial prefrontal cortex (VMN). Note that FPM and DM, despite being located medially, are granular areas.

Granular PFC Underlies Working Memory in Primates

Scientists in the late nineteenth century easily figured out the functions of the sensory and motor cortex by observing the behavior of animals and people with damage to these areas—problems seeing, hearing, or responding to touch defined the corresponding areas of the sensory cortex, and deficits in movement control identified the motor cortex. But the function of granular PFC was elusive.

Part of the problem was that most of the early conclusions were based on simply observing the behavioral effects of brain damage. The contemporary idea that one should use rigorous and well-crafted tasks to study behavioral functions of the brain was not well articulated. In the 1920s, Shepherd Franz and Lashley paved the way for more rigorous scientific approaches to understanding brain functions by introducing the use of behaviorist methods to assess the consequences of brain damage, as opposed to mere clinical observation of behavior.

The key breakthrough regarding the functions of granular PFC emerged in the mid-1930s when C. F. Jacobsen examined the effects of PFC lesions in monkeys using what was called a *delayed response task.* In this task, the monkeys watched as the experimenter placed a food treat under one of two stimuli. A screen was then lowered to prevent the monkey from seeing the stimulus. After a delay, the screen was raised, and the monkey could uncover the treat and eat it. Regardless of whether the delay was a few seconds or several minutes, monkeys without damage to the PFC had no trouble remembering which object the treat was located under. Those with prefrontal damage also performed well with short delays, but were impaired when longer delays were used. Jacobsen described the results by noting that the monkeys with granular PFC damage were unable to hold information in temporary memory for the purpose of solving the problem; instead they could use past learning of what objects are, as well as learned habits.

Jacobsen's studies were the beginning of a long line of research on the role of granular PFC in what we now know as working memory. Particularly important studies in the 1950s by Mortimer Mishkin and Karl Pribram found that the dorsolateral PFC is the area of granular PFC that underlies the monkeys' temporary memory while they perform the delayed response task. Joaquin Fuster and

Patricia Goldman-Rakic later recorded neural activity from individual neurons in the dorsolateral PFC of monkeys in delayed response tasks and showed that some neurons were persistently active during the delay period. This suggested a mechanism by which information might be held in mind by working-memory processes while the monkey decides how to respond.

The dorsolateral PFC is one component of the *lateral PFC,* which also includes the ventrolateral PFC. It is now known that both of these areas contribute to working memory. A third component of the lateral PFC is the lateral orbital PFC, which contributes to working memory in tasks where the value of the stimulus is an important factor. To keep things simple, I will mostly use the broader term lateral PFC, rather than the specific subarea terms, when describing the working memory mechanism.

The contribution of the lateral PFC to working memory in delayed response tasks is illuminated by the fact that it receives inputs from sensory areas of the neocortex. For example, the visual cortex connects with the lateral PFC via two pathways. One connects the visual cortex to temporal lobe circuits that support object recognition, and the other connects it to the parietal lobe circuits that underlie the control of action in relation to external space. These two visual streams converge in the lateral PFC and help bind distinct features (what and where) into complex perceptions that then control actions.

The lateral PFC integrates sensory information not only within a single modality, but also between modalities (Figure 19.3). It is therefore referred to as a *convergence zone* that can form modality-independent representations and allows a monkey or human to "know" what something is by the way it looks, smells, sounds, feels and / or tastes. But the lateral PFC also receives inputs from other convergence zones in the temporal and parietal lobes, allowing even more abstract knowledge about the world, especially areas involved in memory that connect with the lateral PFC directly and by way

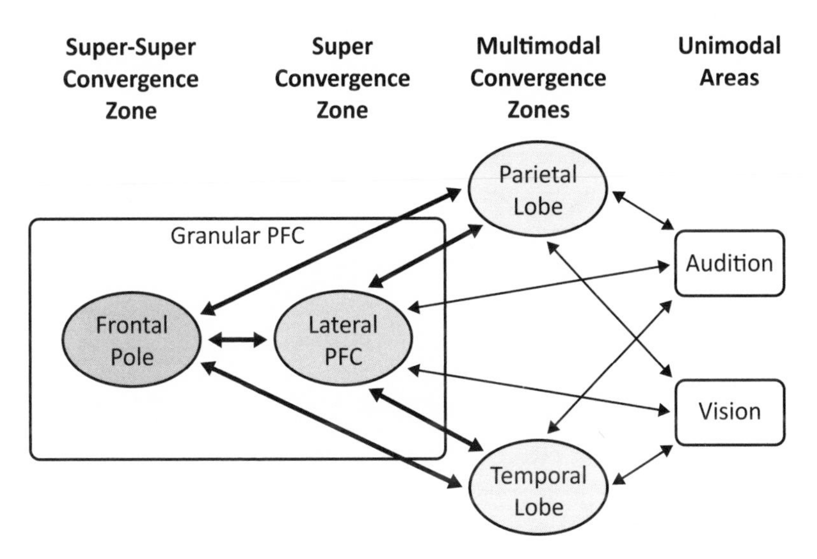

Figure 19.3. Unimodal and multimodal inputs to the granular prefrontal cortex

of indirect connections to sub-granular meso-cortical prefrontal areas. Because of its input from multimodal convergence zones, the lateral PFC is thought of as a *super-convergence zone.*

Lateral PFC areas also connect back to their input areas, allowing top-down control of information processing in sensory and memory systems, and in convergence zones. Because the anterior premotor cortex (APM in Figure 19.2) is part of the granular lateral PFC, and is connected with the dorso-lateral area, information integration within the lateral PFC can be used to control behavior.

Monkey studies described earlier suggested that persistent neural activity in the lateral PFC reflects temporary storage of information when problem solving. With the emergence of Baddeley's model of working memory, lateral PFC came to be thought of as a crucial substrate of both temporary storage and executive functions of working memory. But it later became apparent that persistent activity also occurs in other areas during working-memory tasks, especially sensory and memory areas of the cortex. This led Bradley Postle to suggest that working memory is not simply localized to the lateral PFC, but instead involves a coalition of areas that are controlled by executive functions of the lateral PFC. Postle summarized several specific kinds of interactions between the lateral PFC and other areas that underlie decision-making during the delay period, including monitoring, attending, selecting information, and evaluating expectations about needs, desires, and goals.

Earl Miller and colleagues have evaluated the cellular basis of such processes, and showed that neurons in specific cell layers of the lateral granular PFC exercise top-down executive control over posterior sensory and memory process, regulating their selection for entry into working memory.

Note that the lateral PFC is not the exclusive home of executive functions such as attention. An important role is also played by areas of the parietal cortex. The lateral PFC and the parietal cortex are, in fact, key collaborators in a fronto-parietal attention network, which will come up again shortly.

Some researchers argue that the prelimbic cortex in rats is the homolog of granular PFC in primates. Part of their argument for equating the rat prelimbic PFC with the primate granular lateral PFC is that damage to the prelimbic PFC prevents rats from temporarily maintaining information in working memory while deciding how to act. This argument, however, is problematic. For one thing, it conflates working memory with the granular PFC—that is, it assumes that if rats have working memory, they must have granular PFC. Although rats have a form of working memory, they do not have granular PFC and do not have the kind of working memory that granular PFC supports. Also of note is that rats and primates both possess a prelimbic cortex, and in both it is part of a fronto-parietal attention network that contributes to working memory. The difference is that, in primates, the prelimbic and parietal cortex both connect with the lateral granular PFC working-memory circuits, whereas in rodents, there is no granular PFC to connect with and reap the cognitive benefits it adds. Leading authorities mentioned earlier conclude that the prelimbic cortex in rats is simply the homolog of prelimbic cortex in primates, and is not the rat precursor of granular PFC.

I have emphasized lateral PFC so far. But a second major division of the granular PFC, the *frontal pole,* is also very important in cognition. Unlike the lateral PFC, it receives inputs only from convergence zones, including both the lateral PFC and convergence zones within the parietal and temporal lobes that project to the lateral PFC. Moreover, the frontal pole projects back to all of its convergence zone input areas. It sits at the top of the PFC cognitive processing hierarchy (Figure 19.3) and has been described as the most conceptual part of the brain. Just as the lateral PFC is a super convergence zone, the frontal pole can be thought of as a *super-super-convergence zone.*

Understanding the role of the granular PFC in cognition is a work in progress. But this does not mean that progress has not been made. Earl Miller and Jonathan Cohen offer this cogent summary of the granular PFC's contribution to working memory:

> One of the fundamental mysteries of neuroscience is how coordinated, purposeful behavior arises from the distributed activity of billions of neurons in the brain. Simple behaviors can rely on relatively straightforward interactions between the brain's input and output systems. Animals with fewer than a hundred thousand neurons (in the human brain there are 100 billion or more neurons) can approach food and avoid predators. For animals with larger brains, behavior is more flexible. But flexibility carries a cost: Although our elaborate sensory and motor systems provide detailed information about the external world and make available a large repertoire of actions,

this introduces greater potential for interference and confusion. The richer information we have about the world and the greater number of options for behavior require appropriate attentional, decision-making, and coordinative functions, lest uncertainty prevail. To deal with this multitude of possibilities and to curtail confusion, we have evolved mechanisms that coordinate lower-level sensory and motor processes along a common theme, an internal goal. This ability for cognitive control no doubt involves neural circuitry that extends over much of the brain, but it is commonly held that the prefrontal cortex (PFC) is particularly important.

Brain Mechanisms for Model-Based Goal-Directed Behavior in Primates

As in non-primate mammals, working memory in primates plays a critical role in goal-directed behavior. But the capacities in primates are enhanced in two respects. One is that areas shared between lower (non-primate) mammals and primates continued to evolve after primates diverged. Second, and most important, primates, but not lower mammals, possess granular PFC areas.

The basic goal-directed circuitry is quite similar in rodents and primates. For example, the dorsal medial striatum, sensory-motor cortex, and midbrain dopamine neurons underlie the formation of the instrumental response-reinforcer association; the basolateral amygdala and ventral striatum (nucleus accumbens) contribute to the Pavlovian conditioning of the incentive value of stimuli that predict the reinforcer; the orbital PFC and amygdala retrieve the incentive value; the insula PFC and the amygdala assesses the present value of the reinforcer in relation to homeostatic states; and the hippocampus not only uses sensory inputs from the neocortex to create spatial maps, but also engages semantic memory to locate objects in space and uses information from the orbital and insula PFCs and the amygdala to assign value to objects encountered when foraging. The anterior cingulate plays a key role in resolving conflicts among options and selecting which action to perform.

There are, however, important differences between the rodent and primate systems. For one thing, all of the areas continued to evolve after primates diverged from other mammals—they are, in short, souped up. For example, the meso-cortical PFC areas are more elaborately wired internally and externally, which means that the orbital and insula can handle a more complex integration of information about stimulus and response values and body states; the ventromedial PFC is better able to integrate memories to form schema; the anterior cingulate is more capable of flexibly choosing among options for action as situations change; and the hippocampus in primates not only receives information about stimulus and response values from the amygdala and orbital and insula cortex, but also encodes value itself, forming *value maps* that represent paths through physical space in terms of the expected value of cues and their locations.

But the most important difference involves granular PFC areas. Convergent inputs to lateral PFC working memory circuits allow primates to create considerably more complex representations (mental models) within which higher-order inferences can be made. It's not that rats and other lower mammals completely lack mental models and abstract thought. It is instead that the PFC in lower mammals lacks the degree of sophisticated cognitive centralization and connectivity that the primate granular cortex has. This puts rats and monkeys in different leagues when it comes to mental modeling and abstraction.

The frontal pole adds tremendously to the mix by interacting with other convergence zones, including the lateral PFC. Farshad Mansouri, Etienne Koechlin, and their colleagues summarize the separate contributions of the lateral PFC and the frontal pole in monkey this way: the lateral PFC and its connections make it possible for monkeys to focus on the sustained pursuit of a specific goal. This is optimal in situations where the benefits and risks are fairly stable. But in many situations in life, the benefits and risks are volatile. When this is the case, the ability to disengage from the pursuit of a current goal and consider alternatives can be quite advantageous, especially given that the monkey can maintain the current goal in working memory and return to it if the alternatives do not seem better. The frontal pole of monkeys makes possible this kind of multitasking, which is essential to planning and abstract thought.

Is the Human PFC Special?

A key controversy has been whether humans have any cortical areas that other primates lack. Language areas were long thought to be unique expansions in convergence zones of the human brain. More recently, though, homologs of these areas have been found in other primates. Given that there is no convincing evidence of natural language in other primates, including apes, it is assumed that the convergence zones in question were co-opted for language as humans diverged from ape ancestors.

It has also long been assumed that the granular PFC grew significantly in size when early humans separated from ape ancestors. Starting in the late 1990s this idea was challenged by researchers who argued that the human PFC was about the size expected for a primate of our body weight. Other scholars, like Todd Preuss and Steven Wise, have found support for the traditional view, noting that the newer perspective ignores important differences that are revealed by a finer grain of analysis. They argue that the sensory and motor areas of the neocortex are indeed roughly the same size in humans and other primates when adjusted for overall body size. But they say that convergence zones increased significantly in size. And of these, increases in the granular PFC were the greatest. Furthermore, granular PFC areas have greater connectivity with posterior convergence zones in humans than in other primates, and in humans are the locus of novel gene expression related to energy metabolism.

As noted earlier, the frontal pole is thought to underlie the highest level of conceptual processing in the human brain, especially the ability to engage in hierarchical relational reasoning. In both monkeys and humans, the frontal pole has a lateral and medial component. But evidence suggests that the lateral frontal pole in humans has a component that is lacking in other primates. Just as primates added the lateral PFC and frontal pole, which together allow monkeys to temporarily suspend work on the current goal to consider multiple alternative goals, we humans use our unique lateral frontal-pole component for a more complicated variation: we can continue working on the current goal while considering other possible solution strategies.

Indeed, research by Karin Roelofs in humans has shown engagement of the lateral frontal pole in complex emotion regulation tasks that involve multiple strategies of action control. Using structural and functional connectivity fingerprinting method, they found that the anatomically defined portion of the frontal pole that lacks a homologous counterpart in monkey brains contributes to this task via connections with the sensorimotor cortex, posterior cortical convergence zones (especially the posterior parietal cortex), and the amygdala.

Another idea about the human frontal pole is the *gateway hypothesis* by Paul Burgess and colleagues. They proposed that the lateral and medial parts of the frontal pole make distinct contributions to human cognition by regulating how attention is paid to external stimuli (medial frontal pole) versus internal mental states (lateral frontal pole). The balance between these controls provides a gateway between two of our most basic modes of thought—one about the external world and the other about our inner experiences.

The addition of language and the expansion of granular PFC makes human working memory especially powerful. But a conceptual advance was needed to fully appreciate one of its most important features. As discussed earlier, Baddeley's idea—that the working-memory executive was responsible not just for monitoring and control processes, but also for temporary maintenance—was discarded in favor of the notion that the posterior sensory and memory areas do the lion's share of this temporary maintenance. But it turned out that the specialized systems were insufficient to account for findings from studies of humans working on complex cognitive tasks that require multi-feature binding, especially between modalities (what something looks and sounds like, and where and when you are seeing it). This cannot be done without some temporary storage mechanism that allows integration across specialized systems.

Baddeley suggested a solution to this problem, proposing the existence of an *episodic buffer*—a limited-capacity temporary storage system that might bind diverse kinds of temporarily stored information from subsidiary systems to form a single representation that could then be stored in one's episodic long-term memory. Later studies found that unimodal features are often bound together in unimodal systems, whereas the more complex representations involving multiple modalities are bound in a way consistent with the episodic buffer hypothesis.

The neural basis of Baddeley's episodic buffer is not well understood. But research by Daniel Schacter and Donna Rosa Addis found that when participants were required to integrate stimuli into a complex unit, the frontal pole was activated, seemingly binding the diverse elements of a mental simulation into an episode. The frontal pole may make key contributions to the uniqueness of human episodic cognition.

Working memory had long been associated with conscious awareness. But Baddeley put the icing on the working-memory consciousness cake when he proposed that the episodic buffer is the cognitive gateway to this awareness, a topic that we will revisit in Part V.

What about the Great Apes?

Preuss and Wise have added a caveat to the idea of a lateral frontal pole component that is a unique human specialization, arguing that the possibility that great apes may also have this feature cannot be ruled out. Most of the detailed anatomical comparisons in the literature have been between humans and monkeys. Because little research has been done on chimpanzees due to ethical considerations, we can't be certain that they, or other great apes, lack similar specializations in their lateral frontal pole.

The search for macroscopic unique areas, while useful when successful, is being eclipsed by new technologies that can detect microscopic similarities and differences in gene expression and its effects on the organization and physiological properties of neurons. For example, new findings suggest that gene expression differences have modified the structural organization and physiological properties of the human neocortex, and likely have contributed to behavioral cognitive differences between humans and other primates, including other apes. Clearly, our brain specializations, when combined with human versions of the standard fare of primate goal-directed behavioral control circuits, give us unparalleled capacities for the use of working memory and its executive functions in creating internal models that allow us to forage in our minds for information stored in memory, and to construct scenarios about the future. But remember that "special" means different, not necessarily better. Evolution simply selects what works—what is adaptive. It tinkers and adapts blindly.

There's another reason to suppress any smug feelings about our capacities and the brain mechanisms that make them possible. Preuss has pointed out that because of genetic and molecular changes related to energy metabolism, our souped-up PFC and interconnected convergence zones are "rewired and running hot." In other words, high metabolic rates in PFC may be a double-edged sword. It not only makes our kind of cognition possible. It also may significantly contribute to the human propensity for disorders like autism and schizophrenia, which are associated with dysfunctions of PFC and its cognitive capacities. Living on the edge has consequences.

Summarizing the Neural Basis of Human Cognition

Let me take a moment to consider broadly the various neural connections and cognitive processes involved in constructing and using mental models in behavioral control. As events unfold, complex, parallel interactions between the sensory cortex, memory cortex, meso-cortical PFC, and granular PFC (including the lateral PFC and the frontal pole) shape the content of the mental model. Each process is continuously in flux—that is, updated dynamically in real time.

1. Connections between the sensory cortex and the granular PFC (especially the lateral PFC) allow continuous monitoring and testing of whether sensory processing is about the real world or about internal images (so-called *perceptual reality monitoring*).
2. Connections between the sensory cortex and the temporal lobe activate relevant memories, allowing sensation and semantic memory to be integrated to form meaningful perceptions in the temporal lobe, where schema construction begins in the hippocampus. These memory circuits also connect with lateral PFC.
3. The ventromedial PFC monitors memory accuracy, guides schema construction in the hippocampus, and generates scripts for thought and action based on schema.
4. Connections between the ventromedial PFC and the lateral PFC allow schema and scripts to be incorporated into a mental model that conceptualizes and makes predictions about the situation.
5. Connections between the ventromedial PFC with other meso-cortical areas (insula, orbital, anterior cingulate), and with subcortical goal-processing circuits (in the amygdala and ventral basal ganglia),

allow homeostatic needs and the value of current goals to be integrated into the mental model as it exerts control over goal-directed behaviors.

6. Connections between the lateral PFC and the frontal pole help to determine whether internal or external processing is engaged in the moment.

7. Connections between the lateral PFC and downstream areas—including the posterior parietal cortex, sensorimotor cortex, and amygdala—control the execution of actions based on emotional states.

8. Connections between the lateral PFC and premotor cortex control the execution of model-based behaviors.

9. Connections between the lateral PFC, frontal pole, and mesocortical PFC areas support the use of flexible multitasking, as well as hierarchical, recursive inferential reasoning, when pursuing goals. The result is the ability to behave differently not only because your situation has changed, but also because you have changed your mind.

These processes, including processing within the mental model, I assume, can take place non-consciously. When that is the case, the *mere cognitive realm* is engaged. And next I will offer my views on how non-conscious processes of the cognitive realm contribute to conscious experiences, and the conscious control of behavior.

PART V

THE CONSCIOUS REALM

20

Is Consciousness Mysterious?

Some say consciousness is mysterious. A Google search of *mystery of consciousness,* for example, results in countless hits. For the sake of argument, let's accept that there is some mystery to it. But is consciousness mysterious in the way that the idea of a soul that survives death is, or in the way that the mechanism of evolutionary inheritance was before DNA was discovered?

Philosophy's Hard Problem

Consciousness is traditionally the bailiwick of philosophers. In fact, it is often said that the way we currently conceptualize consciousness comes from Descartes's philosophy, although Greeks covered much of the same mental territory with their own terms, like *soul* (*psyche*). What Descartes did was to emphasize the special quality of consciousness that allows you to know that you are you by virtue of being able to think about who you are. John Locke, as we've seen, later introduced the idea that conscious knowledge of who you were in the past, in the form of memory, allows you to know that you are the same conscious person now. Still later, Alexander Bain offered an even more intimate view of consciousness, saying that we view our thoughts with a "warm eye" and "tenderness."

William James, in the late nineteenth century, similarly referred to the personal quality of conscious states in terms of their "warmth and intimacy."

Today, this special quality of consciousness is front and center in debates about what consciousness is, thanks in large part to the philosopher Thomas Nagel. He describes this quality as a *phenomenal feeling*—the thing that makes red seem red, and fear feel fearful. Therefore conscious states feel like something, whereas unconscious states are accompanied by no such feeling. But why do conscious mental states have this signature phenomenal feeling? David Chalmers, another philosopher, has dubbed this question the "hard problem" of consciousness. Philosophers like Nagel and Chalmers, both of whom are colleagues of mine at NYU, refer to such feelings as *qualia.* Neither believes that consciousness can be explained scientifically. Chalmers accepts that the brain is necessary for experience, but believes that how that happens will forever remain a mystery. Owen Flanagan has dubbed him a "new mysterian," referring to an older dualist tradition that viewed consciousness as dependent on mysterious, non-physical laws.

Scientists have learned quite a bit from philosophers about how to conceptualize complex aspects of the mind for the purpose of guiding research. But the goals of science are often different from those of philosophy. That's why some scientists, myself included, believe that the reason Chalmers's "hard problem" seems so difficult scientifically is because it is a philosophical, not a scientific, problem.

Indeed, I believe that some, maybe most, of the mystery of consciousness results from the current hot scientific pursuit of Chalmer's hard problem. As Anil Seth notes in his book *Being You,* scientists should focus on the so-called real problem of consciousness by doing what scientists do: "explain, predict, and control the phenomenological properties of conscious experience." Until we accept that consciousness is part of our physical, biological makeup, we'll continue to be mystified about what it is.

The Limits of Logic

Philosophers, at least "analytic philosophers" interested in the mind, typically use logic to construct as close to ironclad arguments as possible. Good logical arguments are very pleasing to the human mind—they feel true and compelling. They can, in fact, feel so right that the claims argued for seem like they must be true. In the case of claims about qualia, they must, it seems, point to something real. Although non-physical consciousness is real to non-physicalist philosophers, who are essentially dualists, to scientists, non-physical consciousness is, or should be, an oxymoron. As the physicalist philosopher J. J. C. Smart has pointed out, there is no conceivable scientific experiment that could help one decide about brain-based, materialistic views of mind versus dualistic ones.

William James was both a philosopher and a scientist, and wrote about mental topics wearing both hats, sometimes at once. Describing his inner conflict about the mind as an immaterial soul, on the one hand, and as a physical aspect of the brain, on the other, he lamented, "our reasonings have not established the non-existence of the Soul; they have only proved its superfluity for scientific purposes."

What I take from James's dilemma is that when we turn to philosophy for guidance about the nature of the mind, we need to ask ourselves: is the philosopher making a reasoned argument about a philosophical question, or about how the mind, via the brain, might actually work physically? Qualia seem to be more of a philosophical problem, and, as James said about the Soul, may well be superfluous for scientific purposes, at least as currently conceived.

I don't mean to imply that philosophical theories have no place in science. The philosopher Maurice Merleau-Ponty said, "A science without philosophy would literally not know what it was talking about." That's a bit extreme, but I do believe in a milder version. It goes like this. Scientists are professional collectors and analyzers of data, whereas philosophers are professional thinkers—which means that sometimes scientists can benefit from the logical analyses that philosophers bring to scientific problems. For this reason, I collaborate with philosophers and value their contributions.

But just because the brain can think logically does not mean that evolution wired the brain to be primarily a logic machine. Channeling James, the philosopher U. T. Place noted that "the thesis that consciousness is a process in the brain cannot be dismissed on logical grounds." In the final analysis, while logic is very useful, it is not the answer to every conundrum in life. Daniel Kahneman and Amos Tversky, for example, famously showed that people often rely on heuristics (mental shortcuts), such as referring to what common sense tells them that they or others typically do in similar situations, as opposed to rigorous logic.

Biology is messy, and ultimately scientists need to know when to use, and when to avoid, philosophical wisdom. If we aren't careful about the kinds of ideas we bring to the scientific table from other fields, we risk spinning our scientific wheels on what Cameron Brick and colleagues have called *illusory essences.* As mentioned in Part I, these are concepts or labels or category names that take on explanatory power, as if they were real entities. Self, personality, and intelligence are common examples. Qualia may be another.

A commitment to a non-physical (say, dualist) view of consciousness is perfectly fine as a philosophical position, but it's a deal-breaker as a scientific starting point. If consciousness is not a physical state, it cannot be researched scientifically. The physicalist philosopher Daniel Dennett went so far as to say that Chalmers's hard problem is on par with arguing for the validity of vitalism, the discredited idea in biology that life depends on non-physical elements (see Part II). While Chalmers responded with a philosophical counterargument, Dennett's point is not completely off base. When non-physical explanations of scientific phenomena are taken seriously, and used as a scientific solution to a philosophical conception, we end up in a proverbial explanatory rabbit hole.

For example, integrated information theory (IIT), mentioned in Part II, proposes that consciousness is spread throughout the universe, and is viewed as kind of panpsychism. Such ideas are enjoying a wave of popularity as part of the current revival of secular spiritualism. But IIT is also the subject of damning scientific criticisms. Some say it requires a scientific leap of faith, and others dismiss it as pseudoscience. In 2019, a large group of leading consciousness researchers from various theoretical camps signed a document that argued for tempering enthusiasm for panpsychism on practical grounds:

> Whereas theories of consciousness are of utmost importance for driving further empirical progress, it is important to distinguish empirically productive hypotheses from mysterious and untestable claims such as, for example, the panpsychist view that an inactive set of logic gates could be conscious . . . with limited funding resources, we should be careful about priorities.

Hakwan Lau and Matthias Michel have proposed that we patiently endure non-materialist views of the mind and wait for them to naturally erode as we develop better empirical theories of consciousness. But I think a more immediate scientific solution to the hard problem is also available. And that is simply to accept that science is about, and only about, physical phenomena of the material world. Consciousness would still be hard to figure out, but in a physicalist, potentially soluble scientific way, rather than in a mysterious, unsolvable way.

Science and the Soul

In 1998 I attended a meeting in the hills outside of Kraków, Poland, that was organized by Protestant theologians, sponsored by the Vatican, and included physicists, philosophers, and brain scientists.

The theologians who organized the meeting had the idea that the soul, in the theological sense, equals consciousness, in the scientific sense, and that both are a physical manifestation of the brain of a living human. Given this, the goal was to find a new kind of physics that would allow one's soul to persist as a physical entity after death, and at the time of the resurrection be reunited with its body. In other words, when confronted with a theory that is not compatible with physics, they wanted to find a new kind of physics.

Even among people who accept that consciousness is a physical thing that dies with the body, some have the sense it will never be figured out—it's too hard. But if Alexander Fleming had concluded that bacterial infections were too mysterious and hard to bother with finding a treatment, he would not have discovered penicillin. Or if Watson, Crick, and Franklin had decided that unlocking the secret of inheritance was too mysterious and hard, they would not have discovered DNA. Many more people would have died of COVID-19 if the difficult task of developing a vaccine quickly had not been attempted. Without these and many other solutions to difficult scientific problems, we'd be living in a very different world today.

The solution to a scientific problem is, in fact, always somewhat mysterious until it is found. But that is altogether different from trying to understand scientifically something that lacks a physical explanation. Unless consciousness is a phenomenon of the physical world, we cannot study it scientifically. End of discussion.

What Would a Physical Explanation of Consciousness Entail?

In the sense used so far, a physicalist explanation of consciousness is any scientific explanation that does not violate the laws of physics. But some believe something stronger—that consciousness will ultimately be explained in terms of the laws of physics. This second meaning is an offshoot of a movement that sought to unify all sciences within a single framework. The initiator of this *unity of science* idea was the philosopher Rudolf Carnap, who, in the 1920s, proposed that across all scientific disciplines, "empirical statements can be expressed in a single language"—the language of physics. The core idea of the movement was that physics is more basic than chemistry, and chemistry more basic than biology. Therefore, biology will ultimately be explained in terms of chemistry, and chemistry in terms of physics.

If we take this scheme further, psychology is reducible to brain science, brain science to biology, biology to biochemistry, biochemistry to physical chemistry, and physical chemistry to physics. Ultimately, then, a conscious mental state would be accounted for by physical phenomena such as electrons spinning around their nucleus. If so, then it should be possible to make machines conscious, a topic I will return to later, at the very end.

Theories of consciousness based on physical principles currently do exist. But my sense is that physics is not the best scientific approach to consciousness, at least not at the present time. We would need to know more about the biology of consciousness to relate it to chemistry, and then to physics. For that to happen, we would first need to achieve a neuroscientific understanding of the psychology of consciousness, since we can't reduce something to something else if we can't define it in the framework in which it was conceived.

William Bechtel has recently argued that reduction in neuroscience should be about mechanisms. The question is where does a mechanistic explanation bottom out? If I were a reductionist, I would put my money on a biological account being sufficient, at least for understanding our four realms and their interactions. And even if it is not sufficient, it would be necessary for diving deeper into explanations involving chemical and physical mechanisms.

Yet a physicalist account of consciousness doesn't have to involve reduction to physics. An account based on "non-reductive" interactions between processes at different levels—in particular, interactions between processes in our biological, neurobiological, cognitive, and conscious realms of existence—would, in the end, be a physicalist account.

Reducing Psychology to Behavior

Scientific reduction, as touted in the unity of science movement, was about subsuming higher levels within lower ones. But in the early twentieth century, even before Carnap's ideas were laid out, psychologists produced their own version of reduction when behaviorists staged an internal coup to combat the rampant practice of use of consciousness to explain animal and human behavior. What had usually been explained by reference to consciousness was replaced with behavioral (stimulus-response) explanations.

This exorcism of consciousness was an attempt to make psychology into a "hard" science, a move that received considerable, if late, support from the philosopher Gilbert Ryle, who in 1949 famously said that consciousness is a "ghost in the machine," and therefore has no place in physicalist accounts of human behavior.

In 1993 Daniel Dennett proclaimed that the cognitive movement had triumphed over the barren age of behaviorism, as if the influence of behaviorism was finally over. But that same year, in response to Dennett, Bruce Mangan, another philosopher of the mind, said that behaviorist views were still lingering—that behaviorism, though less vocal, was surprisingly intact. Three decades later, this is still the case in certain corners of psychology and neuroscience, where some continue to complain about the evils of subjective mental states.

Barren behaviorist explanations actually work fine for behaviors of the neurobiological realm (innate or conditioned reflexes, and fixed-reaction patterns, Pavlovian-conditioned responses, and instrumental habits). But they fall flat when applied to behaviors controlled by cognitive mental models and consciousness. If this distinction had been recognized when cognitive psychology was emerging, the two approaches may have been able to coexist, each with its own appropriate share of explanatory power.

Reduction of Psychology to Neural Mechanisms

Having survived behaviorist reduction, psychology then had to cope with attempts to reduce it to neural activity. Neuroscience officially began as a discipline in the late 1960s as a mongrel field consisting of anatomists, physiologists, chemists, developmental biologists, and psychologists. Psychologists were the only group that was not from the so-called hard sciences. Some began to assume that psychology would surely dissolve as the neuroscience underlying psychology matured. The philosophers Paul and Patricia Churchland were major proponents of this view. In the 1980s, they maintained that advances in neuroscience would provide a more precise account of what we call mental states and so eliminate the need for quaint psychological explanations that are typically based on ancient folk concepts.

Consistent with the Churchlands' perspective, Francis Crick, their colleague in San Diego and a discoverer of DNA, said after becoming a neuroscientist that we are "nothing but a pack of neurons," and Richard Dawkins, another prominent biologist, noted that "we are survival machines—robot vehicles blindly programmed to preserve the selfish molecules known as genes." If behavior is indeed determined by such lower-level processes, the quaint folk notion of "free will" will in time vanish into thin air. But that has not happened, and I don't think it will, at least not on the scale proposed by the Churchlands, Crick, and Dawkins.

Folk psychological conceptions underlie our internal mental lives and our external social lives. As the esteemed psychologist Jerome Bruner put it, "folk psychology endows the expected and ordinary events in life with legitimacy and authority." In short, people's commonsense beliefs about minds are a central part of what psychology is all about and will always be central to the endeavor to understand human nature. This is just how the minds of our species work. Yes, over time, folk psychology inevitably incorporates scientific notions. For example, lay people talk freely about pleasure as a surge in dopamine, and love in relation to levels of oxytocin. While both ideas are scientifically dubious, even if we assume they are correct, "pleasure" and "love" are still what is being discussed.

That said, I think that the Churchlands' approach is dead-on when it comes to behavioral control processes of the mere neurobiological realm—that is, model-free neural processes that control reflexes, fixed action patterns, Pavlovian-conditioned responses, and stimulus-response habits. These can, and should, be reconceived in terms that do not imply the involvement of folk-psychological conceptions and conscious mental states in behavioral control. For example, as I discussed earlier, while we often feel "fear" when in danger, or pain when wounded, that does not mean that the subjective feeling of fear or pain directly causes us to flee from danger or writhe when wounded. As the psychologist Melvin Marks pointed out in the 1950s, the use of mental-state words like fear as names for brain states that control defensive behaviors is misleading because it results in the behaviors, and the brain circuits that control them, inheriting the mental-state properties of conscious fear. The solution is to use mental-state words only when referring to mental states, and to use non-subjective state (behaviorist) language for behavioral control processes that are correlated with, but do not depend on, conscious mental states. This is why I say the amygdala circuits that control freezing to danger are not fear circuits. They are defensive survival circuits. The mental state of fear is what happens when you consciously realize that you are in danger.

Now this does not mean that the mental state is not neural. It means that the mental state has to be part of the neural explanation when the behavior is consciously controlled, but should not be part of the explanation when the behavior is not consciously controlled.

Psychology existed without, and could have persisted without, neuroscience. But neuroscience could not exist the way it does today without psychology. The reason that neuroscience is such a popular career choice, and such a popular topic in contemporary culture, is that it helps us to understand how our minds work. And for that, psychological wisdom is needed. One need look no further than the failure of neuroscience-based approaches to the treatment of "mental" disorders. These conditions are named "mental" disorders for a reason.

The Resurrection of Consciousness

I mentioned earlier Karl Lashley's proposal that every conscious state is preceded by non-conscious information processing. This foundational idea allowed the mind to be studied independently of consciousness, and was a key factor in the emergence of the cognitive approach to psychology. That may sound like an example of focusing on the easy part of the problem (non-conscious processes that control behavior), and giving up on the hard part (consciousness). But that is not quite true.

When we study the non-conscious, or to be more precise, the pre-conscious, antecedents of a conscious state, we are microseconds of neural processing away from the conscious experience itself. Much progress has in fact been made by exploring these pre-conscious processes. This does not mean that it will be easy to scientifically cross the precipice into consciousness. But the hurdles we face are not the ephemeral, ghostly qualities of consciousness touted by some philosophers, but instead are the limitations of the tools and conceptions we have today.

Science is a process of building on the past until some discovery creates what Thomas Kuhn famously referred to as a *paradigm shift*. The fact that it has not yet happened in consciousness studies is not a cause for concern—it doesn't happen until it happens.

An early barrier to consciousness research was the depth of antipathy that the behaviorists had toward consciousness as a factor in human psychology. Yet even in the behaviorist era, there were researchers who studied perception, attention, memory, and language in neurological patients and even described the effects of brain damage on conscious experiences. These *neuropsychologists* tended to work in medical settings where they collaborated with neurologists and neurosurgeons, and mostly published in neurological or specialized neuropsychological journals. As a result, they were insulated from restrictions on consciousness imposed by behaviorist dogma.

In the 1980s, the field of *cognitive neuroscience* was created by the neuropsychologist Michael Gazzaniga and the cognitive psychologist George Miller. The core idea was to use the methods of cognitive psychology to understand the psychological functions of the brain. In a sense, cognitive neuroscience was simply a rebranding of neuropsychology. Regardless, it did the trick, because it made cognitive studies of patients a mainstream approach to exploring both the mind (including consciousness) and brain in neuroscience, and at the same time brought the ideas and methods of neuroscience into psychology. Thanks to the fields of neuropsychology and cognitive neuroscience, consciousness is a thriving research topic today.

21

Kinds of Consciousness

The word consciousness is often used as if it refers to a single thing. But it does not. For starters, two broad kinds of consciousness need to be distinguished.

Creature consciousness refers to the condition of being alive, awake, and behaviorally responsive to environmental stimuli. It depends on metabolism for biological existence and on the visceral functions of the nervous system to maintain neurobiological existence. Without these, brain death occurs and cognitive and conscious existence is not sustainable. Creature consciousness is maintained, in part, by an arousal system in the brainstem that allows us to cycle between sleep and wakefulness and that controls our level of alertness when awake. Because creature consciousness is a feature of the neurobiological realm, all animals with nervous systems are conscious in this way, unless they are in a coma or have otherwise suffered severe brain damage. But creature consciousness does not come with mental states.

Mental state consciousness refers to the capacity to experience the world and one's relation to it, and it exists only in animals with conscious realms. It is defined by the content of what one is experiencing, with that content supplied by a variety of brain systems that process information non-consciously, such as sensory, motor, memory, and cognitive systems, among others. Mental state consciousness obviously depends on creature consciousness—you can't be conscious of mental states if you are not alive.

The use of the same word, consciousness, for being alive and responsive to stimuli and for awareness of mental states, is problematic, especially since the modifier *creature* or *mental state* is often omitted. Indeed, in some scientific circles, consciousness is automatically assumed to mean being alive and responsive, and in others it is assumed to mean mental awareness. This can lead to confusion for those not familiar with the difference. So, to be clear, mental state consciousness is the focus here.

Mental State Consciousness

A variety of mental state theories of consciousness have been proposed by philosophers and / or scientists. I will limit the discussion to three broad classes of such theories: first-order, higher-order, and global workspace theories. I emphasize these because they have been studied and discussed extensively, and because most other theories are typically understood as falling within these frameworks. Because most research and theorizing about these and other theories of consciousness have been aimed at sensory, especially visual, consciousness, much of the discussion about the brain mechanisms of mental state consciousness has been focused on the relation between the visual cortex and the PFC, especially granular areas of the lateral PFC.

First-order theory (FOT) is the most basic and the easiest to describe. Theorists from this school argue that you consciously see an apple when a representation of the apple in the world is constructed by your visual cortex. According to this view, the PFC is not necessary for the experience, but it does allow cognitive access to, and reporting about, the visual cortex experience. Ned Block, a philosophical colleague of mine at NYU, is a prominent first-order theorist. Others include Victor Lamme and Rafi Malach.

Higher-order theory (HOT), by contrast, proposes that sensory processing is typically necessary but not sufficient for visual consciousness. In HOT, a subjective visual experience about the external world occurs when visual cortex representations are further processed cognitively by the PFC. This additional processing is often described as cognitive re-representation, or re-description. According to HOT, this re-representation / re-description is said to be crucial for the conscious experience of external events. David Rosenthal, a leading proponent of HOT in philosophy, identified *transitivity* as a key feature of higher-order experience. According to this principle, if one is not aware of being in a state of inner awareness about a stimulus, then one is not having a conscious experience of that stimulus. Another important feature of most versions of HOT is that the higher-order cognitive state is not itself consciously experienced. It is instead a pre-conscious re-representation / re-description that allows the first-order state to be experienced. To consciously experience the higher-order state itself requires additional re-representation / re-description. Variants of HOT will be discussed later in the chapter.

Global workspace theory, originated by the psychologist Bernard Baars, falls somewhere between FOT and HOT. Stanislas Dehaene and colleagues are currently leading proponents, and refer to their model as the neuronal global workspace theory. Like HOT, it posits that visual processing is insufficient and must be further processed cognitively by the PFC, with PFC activity sustained by attention. In contrast to HOT, global workspace theory assumes that PFC activity, rather than creating a re-representation or re-description of the visual information, instead receives and broadcasts lower-order information to widely distributed brain systems involved in memory, cognition, and emotion, among others. Because broadcasting boosts and stabilizes lower-order states, which are responsible for conscious experience in global workspace theory, it is considered to be a variant of FOT.

In sum, traditional FOT says the PFC is not needed for visual consciousness, the global workspace / playground theory argues that the PFC is needed, but mainly for amplifying lower-order states, and the higher-order theory claims that the PFC is crucial for the subjective experience.

Why I Am in the Higher-Order Camp

Richard Brown, Hakwan Lau, and I have argued that HOT has advantages over the other two classes of theories. For one thing, HOT, and only HOT, can account for the transitivity of consciousness—one's awareness of being in a conscious state. In addition, HOT offers a centralized account of conscious experience. By contrast, first-order theories depend on the brain having evolved a separate mechanism for each distinct kind of lower-order conscious experience; this is also true of global workspace theory to the extent that it assumes the conscious experience results from top-down amplification of first-order states. It seems both unlikely and inefficient that the visual, auditory, somatosensory, gustatory, olfactory, motor and memory cortices, emotion networks, and so on, would each have evolved its own mechanism of consciousness. Further complicating the situation is that the cortex of each sensory modality processes multiple features of stimuli. For example, your visual cortex has subsystems that process brightness, color, edges, shapes, and movement. Has a mechanism also evolved for the separate experience of consciousness within each subsystem?

HOT avoids the multiple first-order conscious mechanisms problem by assuming that any and all first-order states are made conscious by a centralized cognitive system involving the PFC. Although global broadcasting depends on the PFC to amplify and sustain lower-order states, unlike in HOT, the PFC itself contributes relatively little to the conscious experience in global theories. Further, HOT more easily accounts for the subjective content of more complex experiences, such as those involving memories and emotions. HOT also offers a way to understand the maladaptive subjective experiences that typify anxiety, depression, and other mental disorders.

Another criticism of first-order theory that is considerably less troubling for HOT comes from findings on what is called *postdictive processing,* where two stimuli several hundred milliseconds apart are perceived as a single stimulus. Studies have shown that diverse patterns of neural activity related to the two stimuli are bundled together into event *chunks.* These can exist in isolation (such as when you casually notice a distant lightning bolt in the night sky), or can be stitched together seamlessly into a more complex meaningful amalgam that unfolds over a longer time period (for example, what you experience and how you react if lightning strikes near you). Local first-order theories of consciousness, which lack cognitive integration of features across sensory and memory systems, cannot easily account for these findings.

A relatively new version of the global workspace theory by Claire Sergent and colleagues adds an interesting twist that accounts for postdictive processing better than local first-order theories do. They propose that a broad network of a lower-order representations are shared and briefly maintained, but only for several hundred of milliseconds, in what they call the *global playground.* This is a kind of mind-wandering mental state, which, as we will see, depends on the so-called default network of the brain. When attention turns to a specific focus, the global playground state enters working-memory via lateral granular PFC circuits, and allows broadcasting in the global workspace.

HOT has much more in common with global first-order workspace theories than local first-order theories. For example, Hakwan Lau, a HOT theorist, and Stanislas Dehaene, a global workspace theorist, proposed that both kinds of processes, especially their conjunction, are typically involved in everyday conscious experiences.

Because not all information that is globally broadcast is consciously experienced, something like higher-order re-representation / re-description is needed to account for the experience itself. And continuous propagation through connection loops, via distal broadcasting, may then be necessary to keep states active long enough for a conscious experience to result from higher-order re-representation, an idea I will build on later.

Michael Graziano's attention schema theory also shares features with the HOT and global workspace theories. Specifically, all three theories emphasize the importance of attention to consciousness, and all three give the PFC a central role. Graziano has, in fact, proposed that his theory unifies HOT and the global workspace theory. But Richard Brown and I have suggested that attention schema, unlike the higher-order states in HOT, do not provide the kind of thought-like awareness required to be conscious of one's mental states. We therefore suggest instead that HOT unifies global workspace and attention schema theory.

Criticisms of Higher-Order Theory

Higher-order theory has its critics. Local first-order theorists say that higher-order representations merely allow cognitive access to conscious first-order states. But that criticism assumes that the first-order state is where the conscious action is, which I just argued against.

HOT has also been criticized for being equated with meta-cognitions. Research on HOT does often use meta-cognitive judgments about one's confidence in their conscious experiences as a proxy for consciousness. But HOT theorists do not typically treat meta-cognition and consciousness as the same. For example, they tend to subscribe to the idea that verbal reports are a more accurate readout of higher-order consciousness than are meta-cognitive confidence judgments. That said, because meta-cognition and higher-order consciousness share neural mechanisms in the PFC, and meta-cognitive approaches offer greater methodological precision than verbal reports in probing the workings of relevant PFC circuits, they are useful despite not being a direct measure of conscious experience. In this regard, Nick Shea and Chris Frith have proposed that if the global workspace theory were to include meta-cognition, it might be better able to account for conscious experience. But a HOT theorist would say that global workspace theory would still need higher-order re-representation / re-description for the conscious experience.

Another criticism is that although people are not necessarily aware of being self-involved in their conscious experiences, higher-order theory assumes that they are. This criticism typically comes up in relation to scientific studies of visual perception, where one has no reason to be particularly involved personally with the meaningless stimuli used. The criticism, however, is based on a misunderstanding of the kind of personal representation that is being called upon in HOT. A full-blown state of reflective awareness in which you are the subject of the experience is not what is in play. Instead, what is usually meant in HOT is a minimal, "thin," or tacit, notion of personal involvement—one that allows you to know that your perceptual states are yours, but without you having to explicitly acknowledge that this is the case. For example, when you enter your home, you don't have to remind yourself that it is your home. You know it is your home because your brain has accumulated knowledge about it from you living there, and you are well acquainted with its features. Similarly, via lifelong mere acquaintance with your biological home (your body) and your psychological home (your mind), you are familiar with what your body and mental states feel like. The importance of this tacit feeling of ownership is most apparent when, through brain damage, it is eliminated. When this happens the person loses their biological and / or psychological ground zero.

Varieties of HOT

Different HOT theorists have various views about what the higher-order state is and how it contributes to consciousness. Lau's theory, called *perceptual reality monitoring*, proposes that consciousness is jointly determined by first-order and higher-order states. In his model, the higher-order state implicitly indexes or points to the relevant first-order sensory information, and distinguishes between sensory states that are real and those that are imaginary or hallucinated. Indexing, in this use, is a kind of sparse low-level re-description that minimizes the need for the higher-order state to have rich content. A related sparse PFC coding view is Stephen Fleming's *higher-order state space (HOSS) theory*, which proposes that the brain uses implicit inferences based on beliefs about learned first-order perceptual representations of the world to construct higher-order states. These two sparse theories bear some relation to the more general notion that concepts are implicit semantic pointers.

Axel Cleeremans's version of HOT is called the *self-organizing metarepresentational* theory. It proposes that consciousness depends on rich higher-order processing based on previous learning drawn from lower-order states. Because the learning involved can be rather simple and automatic, he emphasizes the term "re-description" rather than, or in addition to, re-representation as a process underlying higher-order states. Cleeremans suggests that the universal availability of information in the global workspace is a consequence of these learned higher-order re-descriptions.

Richard Brown's *higher-order representation of a representation (HOROR)* theory, unlike all of the others mentioned, minimizes the essentiality of first-order states—a view supported by findings showing that following damage to the visual cortex one can have a conscious visual experience in the absence of a first-order state. Similar to Rosenthal's foundational HOT, and unlike most other higher-order theorists, Brown emphasizes rich processing based in part on rich memory. But in contrast to Rosenthal, who argues that the higher-order state allows conscious awareness of the first-order state, Brown proposes that the higher-order state itself constitutes the experience. In fact, he says that when there is a mismatch between lower- and higher-order states, one's conscious experience ends up being about the higher-order state.

Brown and I constructed a higher-order theory of emotion based on his HOROR theory. This became the foundation for my *multi-state hierarchical HOT*, which I will describe in the next chapter. But first, I need to set the stage with a little history.

Reinventing the Wheel

Much of the recent debate about the neural basis of mental state consciousness has centered on the question of whether the visual cortex or the PFC is ultimately responsible for conscious experiences. This question came to the fore in the 1990s through the writings of Francis Crick and Christof Koch. Noting that progress in understanding consciousness was stalled, they suggested that because so much was known about the visual system, a focus on visual awareness might be a useful strategy for making headway.

From existing literature, Crick and Koch proposed that connections between the visual cortex and the PFC (by which they meant the lateral, granular PFC) might be important. Specifically, they argued that because the PFC receives input from the late stages of visual processing, but not from the early stages, people are conscious only of visual information received from the late visual stages.

Although the idea that the PFC is important for consciousness was not novel, Crick's fame from his role in the discovery of DNA meant that it grabbed the attention of philosophers, psychologists, neuroscientists, biologists, mathematical modelers, physicists, and the press. The Crick and Koch papers, in short, jump-started contemporary enthusiasm for scientific studies of visual consciousness. The fact is, since the late nineteenth century, consciousness research has mostly focused on vision, and for precisely the reason that Crick and Koch identified. But sometimes it helps to reinvent the wheel.

In the years following the Crick and Koch papers, research accumulated supporting the idea that the lateral PFC has something to do with visual consciousness. For example, when visual stimuli are made consciously imperceptible by subliminal, brief flashes of images, participants typically fail to report seeing anything, whereas with somewhat longer presentations they can report what they saw. And if the brain is imaged during the stimulus presentations, when the participants can accurately report on the stimulus the visual cortex and the lateral PFC are both activated, but when they cannot report, only the visual cortex is active.

Such findings clearly show that activity in the lateral PFC is correlated with conscious experience. Proponents of first-order theories say that such correlations merely demonstrate cognitive access to the conscious first-order state. But other studies, using a procedure called *transcranial magnetic stimulation* to briefly alter activity in the lateral PFC during the visual presentations, show that the participant's ability to report on what was seen is disrupted, suggesting a causal role for PFC.

Rethinking the Sensory-Centric Approach

The sensory-centric view of consciousness has certainly been a productive research focus. But it has had two negative consequences. First, the field got stuck on the idea that the solution to consciousness in the brain was to be found in processing loops between the visual cortex and lateral PFC. And second, most of the research focused on awareness of simple sensory stimuli, such as the color, brightness, shape, or direction of movement of lines, dots, or patches of light. Such basic stimuli offer technical advantages for precise, quantifiable studies of the relation between physical features of stimuli and conscious experiences of those features. But this narrow perspective cannot account for real-life experiences, such as the emotional states we encounter in our minds and talk about when we share our feelings with others, or that we read about in poetry or novels, as when Jane Austin wrote, "You pierce my soul. I am half agony, half hope . . . I have loved none but you." I think a modified higher-order approach is needed, one that accounts not just for awareness of simple sensory stimuli, but also for our memories, emotions, and other personally meaningful experiences, including mental suffering as well.

FOT, HOT, and Our Realms of Existence

From the perspectives of HOT and the three-systems scheme described in Chapter 15, first-order states, being non-cognitive, belong to the neurobiological realm (System 1). By the same logic, higher-order states, being pre-conscious cognitive states, belong to the cognitive realm (System 2). And higher-order conscious states belong to the conscious realm (System 3).

22

Making Consciousness Meaningful

In this chapter, I will explain my version of higher-order theory called the *multi-state hierarchical higher-order theory of consciousness.* It incorporates features of other HOT theories but, unlike most, it emphasizes the central role of memory in making experiences meaningful.

Meaningful Conscious Experiences Require Memory

In daily life, as opposed to the lab, simple visual features, such as shape and color, are typically embedded in the meaningful objects we experience. And those objects are often embedded in meaningful, complex physical or social contexts, called scenes.

The psychologist Jerome Bruner performed a simple but famous study in the 1940s showing how context shapes perception. He presented people with an image that consisted of a column made up of the letters A, B, and C and a row with the numbers 12, 13, and 14. But the B and the 13 in the center of the image were the same visual icon. This was achieved by making the space between the 1 and the 3 smaller than typical so that it could pass as the letter B, but not so small that it could not also pass as the number 13. Participants asked to read the column saw B, and those asked to read the row saw 13.

Meaning comes from memory. We don't innately know about letters and numbers (nor about cats, trees, mountains, pencils, tuna salad sandwiches, or cars). We must learn what these, and other, common things are. And once we do, we use the memories we form to recognize the objects later as members of a particular category, and not members of other categories.

The importance of memory to perception was made clear by the nineteenth-century psychologist and physiologist Hermann von Helmholtz. He claimed that past experiences, stored as memories, allow us to draw "unconscious conclusions" about what is present. He used many examples, including optical illusions and phantom limb symptoms, to support what has since come to be called *unconscious inferences,* or expectations, that shape our conscious perceptions.

Bartlett's schema (see Part IV) can be thought of as memories that enable Helmholtzian unconscious inferences. For example, cherries and red marbles have some similar visual properties (both are reddish, roundish objects that can be similar in size), but from experiences with them we come to know that they are distinct—one is a kind of fruit, and can be eaten, while the other is used in certain games, and is not edible. If you are in a supermarket and see, at a distance, a pile of smallish, roundish, reddish objects in the fruit section, schema activated by the circumstances make you far more likely to infer cherries than marbles. Similarly, a child holding a glass of yellow liquid in a school classroom will be assumed to be drinking something like apple juice, but an adult in a bar with a glass of yellow liquid will more likely be assumed to be drinking beer.

Along these lines, researchers such as Karl Friston, Anil Seth, Chris Frith, and Lisa Feldman Barrett, among others, argue that perception is an expectation or prediction about what is present, based on what we have seen in the past and stored as memory. This sentiment was also expressed by Gerald Edelman, the Nobel Prize–winning immunologist turned neuroscientist, who referred to consciousness as the "remembered present." Richard Thompson, an esteemed memory researcher, put it bluntly: "without memory there can be no mind." And closing the circle, Helmholtz's contemporary, Ewald Hering, wrote:

> Memory connects innumerable single phenomena into a whole, and just as the body would be scattered like dust in countless atoms if the attraction of matter did not hold it together so consciousness—without the connecting power of memory—would fall apart in as many fragments as it contains moments.

Long neglected, interest in the role of memory in consciousness is on the rise.

Flipping the Perspective

I will now begin to construct a theory of consciousness that makes memory a key player in the pre-conscious processing cascade that underlies subjective experiences. In so doing, I will take a page from Crick and Koch's playbook, but with a flip in perspective. Rather than starting with the output connections of the visual cortex to the PFC, I will focus on the PFC and its diverse inputs.

To be clear, brain areas are not isolated modules that carry out functions on their own. Functions depend on interactions between circuits in multiple areas. For example, areas of the PFC contribute to consciousness by virtue of their connections with other areas.

As I discussed in Part IV, the lateral PFC receives inputs not only from unimodal sensory areas, but also from multimodal convergence zones, including memory areas in the temporal and parietal lobes. The memory areas also send inputs to meso-cortical areas of the sub-granular meso-cortical PFC, which both interact with one another and connect with the lateral PFC. Hence the lateral PFC areas get an anatomical double shot of memory—directly from memory areas, and indirectly by way of the meso-cortical PFC. In addition, recall that the frontal pole receives inputs not only from lateral and meso-cortical PFC, but also from temporal and parietal memory convergence zones they interact with.

Memory and other multimodal inputs to the PFC contribute as much, if not more, than unimodal sensory inputs do to complex, real-life conscious experiences. Without memory, sensation is meaningless. Memory is the ultimate curator of complex experiences.

Language processing areas are also multimodal structures. As a result, we can comprehend the same meaning from words that are spoken or written. Imagine you are watching a film in a "foreign" language, and are reading the captions in your native language. If all of a sudden one of the characters starts speaking in your native tongue and the captions disappear, you easily continue to follow the story line and its meaning. Meaning is not tied to the manner in which the words are transmitted (written or spoken). It is instead based on multimodal memories, including schema, that you have acquired about words and their usage.

Earlier I breezed over the role of the meso-cortical areas that receive memory inputs and connect with the lateral and the polar PFC. But one of these sub-granular areas deserves to be highlighted: the ventromedial PFC. The ventromedial PFC is crucial in the assembly of schema, which, as you know by now, are conceptual templates that give meaning to the objects and situations we encounter in life. The construction of schema begins in the hippocampus and other areas of the temporal lobe memory system that connect with the orbital PFC and the anterior cingulate PFC. Outputs of those two areas converge in the ventromedial PFC, which packages schema and shares them with the lateral PFC and the frontal pole, for use in constructing working-memory mental models.

Elements of a Multi-State Hierarchical Theory of Consciousness

My *multi-state hierarchical model of consciousness* replaces the traditional volley between the sensory cortex and the lateral PFC with a more complex anatomical arrangement consisting of a hierarchy of structures, each of which creates different kinds of states that are re-represented / re-described by circuits of sub-granular and granular PFC and that contribute to higher-order mental modeling and conscious experience.

The states that constitute the functional features of the multi-state hierarchical higher-order theory of consciousness, and the brain areas that are associated with these states, include *primary lower-order states* (areas of the sensory cortex); *secondary lower-order states* (memory areas and other convergence zones in the temporal and parietal lobes); *sub-higher-order states* (meso-cortical areas of sub-granular PFC, including the anterior cingulate, orbital, ventromedial, prelimbic, and insula PFC); and higher-order states that re-represent / re-describe / index the various other states to construct mental models in working memory (granular PFC). Also included, but not shown in Figure 22.1, are lower-order states related to goal processing and metabolic needs (basal ganglia, amygdala, nucleus accumbens, hypothalamus), each of which is connected with sub-higher-order areas of the medial meso-cortical PFC.

Two key features of the hierarchical model need to be highlighted. Recall that in Rosenthal's traditional HOT, the higher-order state is a conceptualization that makes the lower-order sensory state conscious. But the traditional view offers no mechanism for memory to support such conceptualizations. My multi-state hierarchical theory includes multiple ways that memory can be used in higher-order

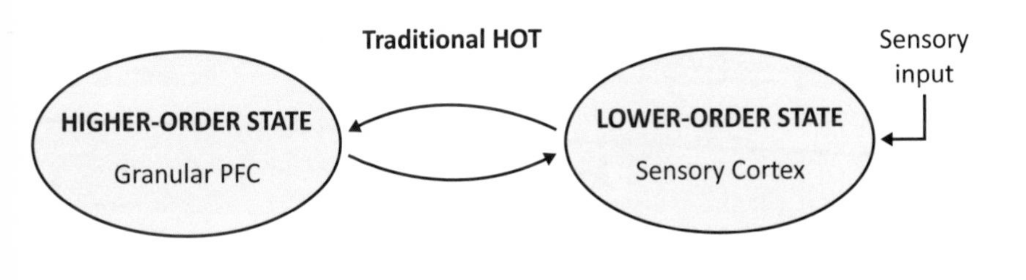

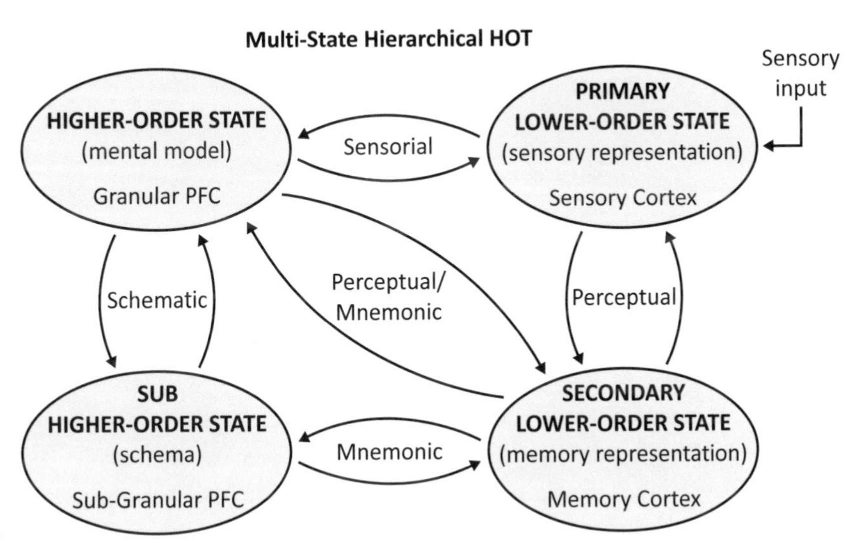

Figure 22.1. Traditional higher-order theories versus my multi-state hierarchical higher-order theory

conceptualizations. In keeping with Richard Brown's HOROR theory, the conceptualizations themselves can be responsible for the experience. That is, although lower-order states often make experience possible, they are not essential.

The philosopher Jorge Morales recently highlighted a significant advantage of my multi-state hierarchical theory for understanding the role of the prefrontal cortex in consciousness. He notes that in my theory, the representations processed in the prefrontal cortex are diverse and redundant and have different degrees of abstraction that reflect their origin from multiple kinds of processing systems (sensory, memory, conceptual). As a result, "the neural focus of awareness may change as an episode develops over time. . . . This feature of the theory prevents a common argument from opponents of HOTs who have critically asked why lesions in the prefrontal cortex do not always result in complete abolition of awareness. . . . LeDoux's answer: the system is dynamic, distributed, and it has many backdoors and backups."

Recurrency, Broadcasting, and Higher-Order Awareness

My theory assumes that higher-order consciousness depends on sustained activity in a working memory mental model. But what sustains that activity? The answer is a suite of recurrent connections—that is, reciprocal connections that form processing loops between brain areas.

Recurrency is hardly a novel idea in consciousness research. For example, first-order theories propose that local recurrency within and between subareas of the visual cortex perpetuates neural activity and enables conscious experiences. Recurrency is also a feature of global broadcasting theories, which propose that conscious experiences depend on sustained activity in reciprocal processing loops between the PFC and a variety of other areas. Axel Cleeremans and colleagues, as noted in the previous chapter, have proposed that recurrency and broadcasting play an important role in enabling and sustaining higher-order consciousness, an idea that I build on here.

In the simplest case, higher-order awareness of a visual stimulus requires distal recurrency between the visual cortex and granular PFC, and local recurrency within both the visual cortex and granular PFC. Attentional control of the neural flow through these loops updates what the visual cortex sends bottom-up to the PFC, and allows the PFC to top-down select what the visual cortex focuses on in the world and what it sends back to the PFC.

Experiences in more typical complex real-life situations tend to be multimodal and involve more complex recurrent interactions. Diverse lower-order areas supply granular and sub-granular PFC areas with various kinds of content (sensorial, perceptual, mnemonic, and so forth). Granular PFC areas re-represent this content and use it to construct a working memory mental model. Local recurrency within both lower-order areas and within the granular and sub-granular PFC temporarily sustains activity within each area. Distal recurrency between the PFC and lower-order areas (including sensory and memory areas) sustains and refreshes the activity of the PFC mental model, and updates its higher-order content to reflect ever-changing lower-order content.

Continuous propagation through local and distal recurrent loops, sustained by PFC broadcasting, may be necessary to keep the overall

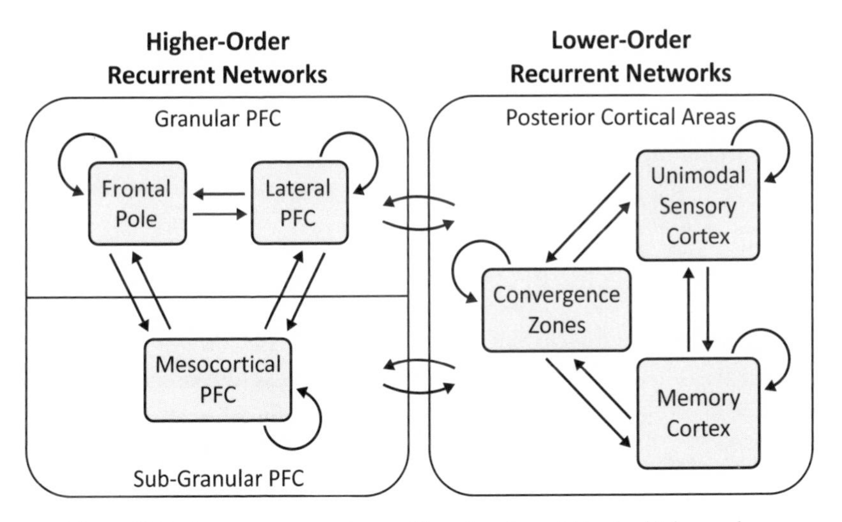

Figure 22.2. Recurrent processing and broadcasting within the higher-order prefrontal cortex, and between the prefrontal cortex and lower-order areas

cognitive network active long enough for a conscious experience to result and persist. But in contrast to the global workspace theory, which associates consciousness with the signal broadcasted from the PFC, in my theory, consciousness depends on recurrent activity within the pre-conscious PFC working-memory mental model, with local and distal broadcasting sustaining propagation through the relevant recurrent circuits. Findings showing that increased functional connectivity occurs within the PFC when working memory is engaged are consistent with the idea that local recurrent activity in the PFC is involved in sustaining the mental model and hence conscious experience.

The PFC mental-model content typically depends on bottom-up inputs from lower-order areas. But a key feature of mental models is that they can use memory (especially schema) to simulate (imagine) present situations. This may account for how patients with visual cortex damage, and therefore without lower-order visual states, can have vivid visual experiences, though not of the external world. I mentioned this earlier in connection with Richard Brown's HOROR theory.

To be clear, my proposal that conscious experience involves a pre-conscious working-memory mental model does not mean that working memory and mental models are sufficient for a conscious experience to occur. More is needed.

Unpacking the PFC's Contributions to Consciousness

I have described recurrent activity within the PFC as if it involves recurrent connections between three broad areas (the lateral, polar, and meso-cortical PFC). This caricature is useful for explaining the basic idea, but it is a gross over-simplification of the actual situation. For starters, each of the three regions is composed of multiple subareas and circuits that are interconnected with one another, and with subareas in other parts of the PFC (see Part IV).

Earlier, I mentioned that stimuli processed non-consciously several hundreds of milliseconds apart in lower-order areas can come to be experienced simultaneously as part of a unified, wholistic *event chunk*. This idea of non-conscious integration of information over time is compatible with the kind of neural framework I just described in which conscious experience is supported by the continuous updating and integration of representations in a non-conscious higher-order mental model.

I will next discuss what each area of the PFC contributes to conscious experiences. In so doing, I will borrow from summaries of the known role of specific PFC areas in cognitive functions by Kalina Christoff, Karin Roelofs, Dagmar Zeithamova, and their colleagues.

The *lateral PFC* makes possible flexible, top-down executive control by interacting directly with lower-order sensory and motor areas, with memory areas and other convergence zones, and with the amygdala. By constraining what enters the higher-order lateral PFC circuits, conscious thought transitions are minimized, allowing a focus on the present goal and its expected value—or, in working memory parlance, *attention focusing* and *refreshing*. The ventro-lateral component focuses on relatively low-level abstract concepts, whereas the dorsolateral is more involved in higher-level abstraction.

The *frontal pole* is thought to contribute to the highest levels of abstract conceptual reasoning and problem solving (see Part IV). It uses memory to assess novelty, and to evaluate alternative goals and predict their values, and it focuses thoughts by suppressing those that are irrelevant to the goals under consideration. Creative thinking is facilitated by loosening such constraints.

Meso-cortical PFC areas (including the ventromedial, prelimbic, anterior cingulate, orbital, and insula PFC) function as conduits that route information related to schema, concepts, past and present goal states, and homeostatic (metabolic) need states, to lateral and polar granular PFC areas.

Misunderstandings about the Prefrontal Cortex and Consciousness

Higher-order, global-workspace, and attention schema theories each call on the prefrontal cortex (PFC) in their account of conscious experience. Supporting the idea of the involvement of the PFC in consciousness are findings showing that increased functional activity in the PFC is correlated with conscious experiences, as well as with performance on meta-cognitive tasks that are thought to be related to consciousness. While these correlations are suggestive, they do not directly implicate the PFC in consciousness. But results of other studies do. For example, people with PFC damage have impairments in consciousness and a decline in performance on meta-cognitive tasks related to conscious experience. While first-order theorists have challenged these conclusions, their challenges overlook important nuances.

The PFC processes information consciously and non-consciously. Therefore, one cannot assume that a state is conscious just because it depends on the PFC. At the same time, just because PFC damage fails to result in a measurable loss of conscious awareness does not mean that the PFC is not involved in the conscious experience. For example, damage to the PFC may alter consciousness in subtle ways without necessarily eliminating it. And much depends on how that loss is assessed. Whether patients with PFC damage are conscious in the same way they were prior to the damage may not be revealed by standard clinical exams, which are relatively crude compared to rigorous scientific tests. Another complication is that verbal report, the gold standard in human consciousness research, depends on granular PFC. That is, when the PFC is damaged, the verbal report can be compromised either due to a change in speech production or because PFC damage also often results in confabulations that compensate and cover for losses.

Because there are many sub-regions of the PFC, it is important to have as much information as possible about the exact locus of damage. Does the damage involve both granular and meso-cortical PFC? Which subareas of granular and/or meso-cortical PFC are damaged? Is the damage to the PFC in one hemisphere or both? How much damage is there outside of the PFC? Because lesions in clinical cases are often messy, it can be difficult to answer these questions precisely.

Functions lost after brain damage are often assumed to reveal what the damaged brain area typically does. But this can be misleading since the remaining undamaged areas of the brain can compensate for the loss. For example, patients with damage to the visual cortex can have conscious visual experiences, if the PFC simulates the missing lower-order state and produces a hallucinated visual experience that is difficult for the person to distinguish from reality. This is consistent with the idea that higher-order top-down processes contribute as much, and maybe more, to perception than does the lower-order state.

What if the visual cortex is intact, but the lateral PFC is damaged? If the frontal pole is not included in the damage, it may compensate. After all, the frontal pole and lateral PFC have many of the same connections. But even when the lateral PFC is present, the frontal pole normally carries the lion's share of conceptualization, with the dorsolateral PFC mainly taking care of lower-level goals and action control. By contrast, if the frontal pole is selectively damaged, which is very uncommon given its location, an intact lateral PFC might be able to pull off a "lite" version of what happens when both are present.

So what happens if both the frontal pole and lateral PFC are completely damaged? The burden would then fall to sub-granular meso-cortical areas of PFC, which instantiate sub-higher-order states. In lower mammals—rodents, for example—meso-cortical PFC circuits construct working-memory mental models in goal-directed behavior. But the meso-cortex is as high-order as it gets in non-primate mammals. Hence, when granular PFC is missing in humans, a less sophisticated form of consciousness based on schema processing and mental modeling capacities of the meso-cortex, one that is normally overshadowed by the more elaborate mental model capacities of granular PFC, may be unmasked. This unmasked meso-cortical PFC consciousness in humans would not be strictly rat-like. Meso-cortical areas of PFC continued to evolve when primates diverged from lower mammals, apes from lower primates, and humans from apes. Moreover, language and verbal

(Continued)

Misunderstandings about the Prefrontal Cortex and Consciousness *(continued)*

memory make the human meso-cortex able to conceive of and create a mental model of both one's inner life and the external world in ways that other mammals with a meso-cortex cannot.

If all of the PFC (sub-granular and granular) were missing, other kinds of consciousness could conceivably be unmasked. For example, although the executive functions of the PFC would be lacking, areas of the parietal cortex also contribute to executive control and are frequently considered a crucial partner with the PFC in higher-order and global workspace theories. In addition, the hippocampus is no slouch when it comes to higher-cognitive processing, and has, in fact, been the subject of theories of consciousness, especially with respect to spatial consciousness, but more generally as well. Perhaps, then, when the PFC is extensively damaged a primitive kind of consciousness dependent on either the parietal cortex or hippocampus, or their interaction, might be revealed.

It is important in this context to mention a paper by Björn Merker in which he used findings from patients with hydrocephaly, and hence very limited cortical tissue, to argue that consciousness does not at all depend on the cerebral cortex. In the studies he cited, consciousness was judged by sensory-motor responsiveness, not by evidence of mental states. This is consistent with his description of the states in question as states of wakefulness and responsiveness to sensory stimuli—in other words, as creature consciousness. Merker clearly stated that he was not referring to the kind of consciousness typically experienced by adult humans. Even so, many who have cited his work, such as the psychiatrist Mark Solms, have treated the findings as evidence for mental-state consciousness without a cortex. But this is misguided, for mental-state consciousness cannot be presumed from primitive sensory-motor processes in the neurobiological realm.

Findings by Stephen Fleming and colleagues are roughly consistent with this general picture. Using meta-cognition paradigms, they implicated meso-cortical PFC circuits in implicit pre-reflective processing, and lateral PFC circuits, especially circuits within the lateral frontal pole, in reflective conscious experiences, and in judicating whether to be truthful or deceitful.

Just because I've placed so much emphasis on PFC processing does not mean that the PFC is absolutely necessary for conscious experiences. There are multiple kinds of conscious states that might be unmasked when the PFC is either partially or even completely damaged. That said, consciousness without a PFC is akin to making a cross-country trip in a covered wagon pulled by oxen, versus a gas-powered, air-conditioned car guided by GPS and outfitted with satellite radio. Both options could get you to your destination. But the trips would be quite different.

Looking Forward

If I were you, I would still be asking, "How does a working memory mental model spawn conscious experience?" The short answer is that the mental model generates a story line, a narration, that creates the content of our conscious experiences. The narration, in other words, is the event that crosses the consciousness finish line. But you'll have to wait patiently until the final chapter for the details, since I need to fill in other pieces of the puzzle first.

23

Fact-Knowing and Self-Knowing Consciousness

I have established the importance of memory to consciousness. But different kinds of memories underlie different kinds of conscious experiences.

Long-term memory is typically divided into explicit and implicit categories. *Explicit memories* are formed and stored in a way that can be retrieved into working memory, where they can be introspected and consciously experienced. Because we can "declare" their contents at will, these are sometimes called *declarative memories.* By contrast, *implicit memories* are stored by way of procedural learning. They can be retrieved and expressed only through nonverbal body responses. In this chapter, I focus on explicit declarative memories and the kinds of consciousness these memories support. Implicit memory is the topic of the next chapter.

Types of Explicit Memory

A key example of explicit, consciously accessible memory is *semantic memory,* which, as we saw in relation to working memory, involves factual and conceptual knowledge, and, importantly, is the foundation of schema. Another kind of explicit memory is more personal.

It is about your very own experiences. For example, you can learn semantic facts about Greece by reading a travel book while sitting at home preparing for a trip there. But it is quite another thing to experience the culture on your own. When you do, you form memories that have specific content. And when retrieved, they include a place stamp about where *you* were, a time stamp about when *you* were there, and an event stamp about what *you* saw and did and who *you* were with. Where, when, and what are essential components of so-called *episodic memories.*

When Endel Tulving first proposed the distinction between semantic and episodic memory in the early 1970s, he treated these as truly different kinds of states. Current understanding holds that while semantic memories can occur separately from episodic memories, episodic memories always include relevant semantic information. But there is a unique element of distinctiveness to episodic memories—they alone allow you to construct a sense of personal identity and continuity in your life.

Martin Conway has discussed some key defining features of episodic memory, which I liberally paraphrase here. Every moment in waking life is an episode, but memories of most episodes are quickly forgotten unless they are especially meaningful to you. Episodic memories code experiences in terms of time, location, and content. Although they can seem quite vivid, vividness and accuracy do not necessarily go hand in hand in episodic memories.

Although traditionally thought of as a literal archive of past experiences, episodic memory is not a carbon copy of what happened in the past. Instead of retrieving exquisite details about the episodes, we use the stored gist and relevant semantic schema about the experience to construct a scenario, a narrative, about what happened, and when and where it occurred. Because the construction process is messy, the memory that is retrieved can deviate, sometimes significantly, from what was experienced, so that two people can have very different memories of the same event.

Explicit Conscious Content Is Enabled by Explicit Memories

A decade or so after introducing the semantic-episodic distinction, Tulving proposed that these two kinds of memories underlie two different kinds of conscious experiences. He referred to these as *noetic* and *autonoetic* consciousness.

Noetic states of consciousness have mental content based on semantic facts and conceptions about the world. They often involve schema, which, as we've seen, provide conceptual templates that give meaning to the objects and situations we encounter in life. States of *autonoetic* consciousness, by contrast, have episodic mental content in which "you" are the entity that had the experience. Tulving therefore proposed that the two kinds of consciousness were associated with distinct kinds of knowledge. Noetic states, he said, come with factual (including conceptual) knowledge, while autonoetic states come with personal knowledge about who you are.

Perhaps the most celebrated feature of autonoetic consciousness is *mental time travel.* When Locke said that, through memory, we know we are the same person today as in the past, he was, in effect, referring to the ability to engage in mental time travel. Similarly, when William James wrote, "Memory requires more than mere dating of a fact. It must be dated in *my* past," he too had something like mental time travel in mind.

In addition to allowing you to construct episodes from your life's past, episodic autonoetic consciousness allows you to mentally experience the present, and to imagine possible situations that may occur in your future life. Paraphrasing the neuroscientist György Buzsáki, episodic memory functions like a search engine that we use to construct the past, present, and future. Autonoetic consciousness is, in effect, a very sophisticated form of mental modeling.

Tulving has long emphasized the importance of consciousness for understanding memory and of memory for understanding consciousness. In the process of writing my 2015 book *Anxious* I became a convert, and ever since have touted these relationships.

Meta-Cognitive Processes Underlying Noesis and Autonoesis

Janet Metcalfe and Lisa Son interpreted Tulving's noetic and autonoetic conscious states in terms of meta-cognitive representations of semantic and episodic memory. Because meta-cognitive representations are higher-order with respect to memory, their model is, in effect, a kind of higher-order theory of consciousness, which they acknowledge.

For example, noetic consciousness of an apple on a tree involves the meta-cognitive (higher-order) representation / re-description of lower-order sensory and memory (including schematic) information in the working memory mental model. This results in a conceptualization, and ultimately in the experience, of the apple and its relation to its situational context. By contrast, an autonoetic experience of the same situation occurs if the apple triggers memories of having picked apples as a child via the meta-cognitive (higher-order) representation / re-description of episodic information in the working memory mental model. When that autonoetic experience transpires, the conscious person becomes the entity experiencing the apple and having thoughts about the experience.

The "Self" in Autonoesis

Explicit consciousness, especially autonoetic consciousness, is often said to involve the "self" via personal episodic memories. But the self-involved in these experiences is not an agent that does things. This self does not take you to your past or future when you mentally time travel. Instead, your cognitive capacity for mental time travel allows you to construct those experiences in which you are the subject. To the extent that a self is involved, it is *informational* (the result of a constructive narrative based on episodic memories) rather than *agential* (an entity that narrates and controls behavior). But regardless of whether one is explicitly aware of their personal involvement in a conscious experience, as I will argue in the next chapter, tacit personal knowledge accompanies it, and accounts for what it feels like for you to be explicitly conscious of your experiences, including both noetic and autonoetic experiences.

Brain Mechanisms Underlying Noesis and Autonoesis

Studying consciousness can be challenging, especially when trying to understand the brain mechanisms involved. But Tulving's theory, and Metcalfe and Son's meta-cognitive twist, together provide a back door that makes it easier than it otherwise would be. Namely, because semantic and episodic memory can be thought of as lower-order states that are cognitively re-represented / re-described in the process of creating noetic and autonoetic conscious states (see Figure 22.1), much of what we need to know to get to the border that separates non-consciousness from consciousness is known from existing research on semantic and episodic memory.

Table 23.1 lists key cortical areas that contain circuits involved in semantic and episodic memory, and shows these in relation to the hierarchical levels of lower- and higher-order states in Figure 22.1. As you can see, semantic and episodic memory are processed by separate, but overlapping, circuits. Areas where semantic and episodic inputs converge are shown in the "Convergent" column.

The neural basis of autonoesis itself, as opposed to the neural basis of the episodic memories underlying autonoesis, was first described in detail in 1997 by Mark Wheeler, Donald Stuss, and Tulving. They made a strong case for the involvement of PFC areas, drawing on classic findings by Alexander Luria that led him to conclude that following frontal damage, patients lost a critical attitude toward

Table 23.1. Key brain regions involved in semantic and episodic memory, and their relation to hierarchical states

Representational States	*Semantic Memory*	*Convergent*	*Episodic Memory*
Primary Lower-Order (sensory)	**Sensory Neocortex** (ventral visual stream)		**Sensory Neocortex** (dorsal visual stream)
Secondary Lower-Order (sensory & memory)	**Neocortical Temporal Lobe** (inferior, superior, polar)		**Neocortical Parietal Lobe** (precuneus & retrosplenial)
	Medial Temporal Lobe (perirhinal, lateral entorhinal)	**Medial Temporal Lobe** (hippocampus)	**Medial Temporal Lobe** (parahippocampal, medial entorhinal)
Sub-Higher-Order (memory & schema)	**Meso-Cortical PFC** (orbital, medial)	**Meso-Cortical PFC** (insula, medial, ventromedial)	**Meso-Cortical PFC** (anterior cingulate)
Higher-Order Circuits (mental models)	**Neocortical PFC** (dorso- & ventro lateral; lateral & medial polar)	**Neocortical PFC** (dorso- & ventro lateral; lateral & medial polar)	**Neocortical PFC** (dorso- & ventro lateral; lateral & medial polar)

themselves. They also used findings from patients who had undergone frontal "psychosurgery," who often discussed their problems in a detached manner, as if they were a casual observer. Both types of patients seemed to have little insight into their problems. Wheeler, Stuss, and Tulving interpreted these and other findings to mean that frontal damage has devastating effects on the ability to reflect about who one is in a meaningful way because of the loss of a fully functioning capacity for autonoetic consciousness.

As we have seen, consciousness is not solely a product of a single brain area, or even of circuits in a single area. It takes an interactive collation of circuits in multiple brain areas. Amnon Dafni-Merom and Shahar Arzy recently reviewed the neural basis of autonoesis and presented evidence implicating not only areas of the PFC, but also areas in the temporal and parietal lobes, especially those areas associated with episodic memory. In the scheme I have been developing, autonoesis results from the higher-order representation / re-description of lower-order sensory and memory states and involves PFC circuits, including circuits in both intermediate-order (sub-granular meso-cortical) and higher-order (granular neocortical) areas, as well as their connections with circuits in sensory, memory, and other areas.

Past, Present, and Future Episodes

That Tulving linked autonoetic consciousness to episodic memory has sometimes been interpreted to mean that whenever we are in a state of autonoetic consciousness, we are reconstructing some past event in our life. But episodic memory in the form of schema about past episodes provides templates for constructing the episodic present and imagining possible episodic future scenarios.

Conceptions closely related to, if not the same as, mental time travel to the future include what Randy Buckner and Daniel Carroll have called "self" (personal) projection; Thomas Suddendorf, Adam Bulley, and Jonathan Redshaw have referred to as prospection (future thought and foresight); and Daniel Schacter and Donna Rose Addis have described as constructive episodic future simulation (imagination). What ties these together is the involvement of one's hypothetical future.

Given the close relation of personal-projection, prospection, episodic future thinking, and future mental time travel, they likely share common brain mechanisms. Indeed, these processes each depend on a group of interconnected brain regions that are part of what is commonly referred to as the *default mode network* of the brain. Involved are circuits in the temporal and parietal lobes, and in the meso-cortical prefrontal cortex. This network is thought to be active when the brain is at rest. In this state, the mind is simply wandering, with "you" often the subject of such experiences.

But the default network is not the sole contributor to personal projection, prospection, episodic future simulation, and mental time travel to the future. These states also depend on the lateral PFC for retrieving episodic memories from the temporal and parietal areas and integrating them into the mental model. In addition, the frontal pole enables the use of recursive, hierarchical reasoning when evaluating alternative strategies and goals, and it regulates shifts in perspective from the outer environment to our inner experiences. The contribution of the frontal pole in attending to inner versus outer information is related to the perceptual-reality-monitoring theory of higher-order consciousness, mentioned earlier. The frontal pole is also important for understanding the relation between episodic memory, imagination, and subjective meta-cognitions about one's emotions. While global workspace, attention schema, and higher-order theories each call on similar processes and brain areas, as I noted earlier, I believe that HOT is better able to account for conscious experiences in which you are the subject.

Also related to mental time travel and its various manifestations is *subjective meta-cognition,* which includes thoughts about who you are, thoughts about an episodic memory of something you did in the past, or a prediction about possible future episodes you may have. Another example is the ability to change your mind by considering new information, something that Stephen Fleming and colleagues found involves the frontal pole. They have also shown that the frontal pole, especially the lateral frontal pole, allows you not only to express outwardly what you consciously believe, but also to express alternative positions that are not actually believed, something done when being deceitful, whether for nefarious or altruistic reasons.

Other research has found that one's subjective sense of the intensity of an emotional experience recruits the frontal pole. Still other research shows that transcranial magnetic stimulation of the frontal pole interferes with the use of meta-cognitive self-reflection in making choices. *Theory of mind*—the ability to imagine what is on the mind of another, based on an understanding of how one's own mind works—is a further instance of a subjective meta-cognitive capacity that involves the frontal pole.

Prefrontal Cortex, the Default Network, and Who You Are

The topic of self in the brain has a long history that goes back to at least the famous case of Phineas Gage in the nineteenth century. Due to an accident while working on a railroad, Gage suffered damage to medial areas of his PFC. Once a kind, thoughtful person, he became erratic and unreliable. Wheeler, Stuss, and Tulving's data set mentioned earlier included patients with large brain lesions that involved significant damage to both the lateral and medial PFC areas; damage to the medial PFC no doubt contributed to the deficits in personal awareness that they described. Much contemporary research under the rubric of "self" has, as mentioned earlier, also implicated the medial meso-cortical PFC and other areas of the "mind wandering" default mode network, likely because our thoughts are often about ourselves when our mind wanders.

The term *medial PFC* typically refers to sub-granular meso-cortical areas, including ventromedial, prelimbic, anterior cingulate, and medial orbital cortices. But when it is used in connection with the default network, the dorsomedial PFC and the medial frontal pole areas are also included. This is a problem. Despite being located medially, the latter two are granular areas of PFC (see Figure 19.2). Lumping together primate-unique granular areas with more primitive, sub-granular, medial areas shared by all mammals, without acknowledging the difference, ignores the distinct contributions that the granular and sub-granular areas make to the processing of information about who you are.

Recognizing the difference between granular and sub-granular areas of the default network seems essential to understanding the evolutionary history of the default network and its role in personal processing, as well as in clarifying the nature of interactions between the default network and granular PFC areas. For example, antagonistic interactions between the default network and granular PFC cognitive networks are often viewed as necessary for optimal cognitive performance—that is, things work best if they are active one at a time. But recent studies emphasize that under some conditions, the default network areas cooperate with granular PFC areas. A key question, though, is whether the medial granular PFC areas are more aligned with lateral granular PFC areas traditionally involved in working memory, or whether they are more closely tied to mesocortical areas. One possibility is that the medial location of the medial frontal pole and dorsomedial PFC is primarily a real estate issue, rather than a functional one—that's just where they were squeezed in as the granular cortex evolved. This is an issue that begs to be resolved.

Memory and Consciousness Revisited

Explicit memories are often referred to as conscious memories. That's a bit of a misnomer. They are instead memories that can be consciously experienced when retrieved and appropriately re-represented / re-described in higher-order networks. When this results, depending on whether semantic or episodic memory was engaged, we can have noetic experiences about facts and concepts, or personally meaningful autonoetic experiences. But let's not forget that Tulving also proposed a third kind of conscious experience, anoesis, which we turn to next.

24

Non-Knowing Consciousness

Endel Tulving dubbed conscious states based on implicit procedural memory states of *anoetic consciousness.* And just as he associated autonoetic consciousness with self-knowledge and noetic consciousness with factual knowledge, he associated anoesis with *non-knowing knowledge.*

At first blush, this makes no sense. If procedural memory is implicit—that is, unconscious—and if it is expressed only in behavior, how can its re-representation / re-description result in a state of conscious awareness? And what the heck is non-knowing knowledge anyway?

Given how little Tulving wrote about anoesis, compared to noesis and autonoesis, it seems it was the state he had the least interest in. One possibility, then, is that he simply thew it into the mix to tidy things up. But I doubt that. He is probably the cleverest, most thoughtful psychologist since William James, and he likely was pondering something important, even if he didn't fully articulate it in his writings. Below is my interpretation of what he may have had in mind.

The Anoetic Paradox

The fundamental problem in understanding anoesis is the incompatibility of three ideas: that procedural memory is unconscious; that unconscious procedural memory is the basis of anoetic states; and that anoesis is a form of consciousness. I believe I can make these ideas mesh with one another.

Semantic and episodic memory are tightly constrained psychological conceptions. And each has its own set of neural circuits. But procedural memory is more of a grab-bag notion. It consists of diverse kinds of learning spread all over the brain, and that are connected only by the fact that they are not learned explicitly the way that semantic fact knowledge or episodic personal knowledge is.

Marie Vandekerckhove and Jaak Panksepp related Tulving's anoesis to what William James variably referred to as the *fringe, penumbra,* or *halo* of the *stream of consciousness.* These vague, tacit states, James said, allow explicit, content-filled conscious states to feel "right."

James's fringe and its "feeling of rightness" has been adopted and adapted by some contemporary philosophers. Bruce Mangan, for example, says that rightness is the most important factor that controls conscious / nonconscious interactions. He extended fringe feelings to such present-day psychological research topics as *gut feelings, meta-cognitive confidence judgments, tip-of-the-tongue* experiences, and the *feeling-of-knowing.* To Mangan's list, I would add Leon Festinger's notion of *cognitive dissonance,* which results in a feeling of "wrongness" when inconsistent beliefs sit side by side in the mind, and Daniel Oppenheimer's notion of *fluency,* the vague subjective experience of ease or difficulty associated with completing a mental task.

But long before Mangan, and even Tulving, Arthur Reber introduced the idea of unconscious implicit learning to characterize how one develops intuitive knowledge about the world and uses it to tacitly understand the relation between co-occurring environmental events.

According to Mangan, the tip-of-the-tongue phenomenon was a prime example of the fringe for James. This refers to a situation in which one is, for example, unable to recall someone's name but is confident that they know it. Typically, an entire complex of information about the person can be explicitly known (their age, hair color, ethnicity, personality traits, and so on), despite the inability to come up with their name. Zoltán Dienes has obtained scientific support for the idea that people can be confident of knowing that they know something (it feels familiar) but without knowing what or how they know it.

Consistent with this line of thought, Asher Koriat has referred to such vague fringe feelings as *sheer subjective experiences.* These, he said, depend on implicit inferences shaped by unconscious processes that monitor memory states and yield a present or absent conclusion, rather than the specific content that you feel you know. This unconscious monitoring of memory has been called procedural or implicit meta-cognition. Stephen Fleming has recently made a similar point—that being able to report on whether something is present or absent is different than being able to report about one's awareness of the item in question.

Why does consciousness have a dual organization consisting of a tacit fringe component and an explicit focal condition? Mangan explains this situation as a tradeoff. Through working memory, stimuli can be selected and attended to in focal consciousness, and result in articulatable (introspectable and reportable) content. Fringe states accompany explicit conscious states and imbue them with a feeling of familiarity and rightness, but in doing so make minimal demands on working memory and articulation resources. The function of the fringe, Mangan says, is to bring to bear in a highly condensed way relevant information in the extensive unconscious repository of procedural memory, thereby finessing the limited capacity of focal consciousness.

As I noted earlier, procedural memory is a grab-bag concept that subsumes a diverse range of neural circuits that learn and adapt and that are present in literally all areas of the brain, and even in the spinal cord and peripheral sensory and motor ganglia. To the extent that these procedural learning processes are "automatic" and thought of as "not conscious," they are, by definition, part of what Mangan referred to as the "vast unconscious context." But he also includes some instances of the "cognitive unconscious" in his vast unconscious (this mixing of neurobiological and cognitive realm processes is another example of why the traditional dual-system approach needed an overhaul—see Part IV).

If we take focal, explicit consciousness to encompass noesis and autonoesis, and fringe consciousness to refer to anoesis, the psychological relation of vague anoesis to explicit consciousness becomes clear. Anoetic states are what give noetic and autonoetic states the warmth, tenderness, and intimacy that James, and Alexander Bain before him, talked about. Sheer feelings about the fringe, in other words, would seem to be one and the same as anoetic consciousness.

Metcalfe and Son proposed that, like autonoetic and noetic consciousness, anoesis requires re-representation / re-description, in this case procedural or implicit meta-cognition. But in order for this higher-order account of anoesis to work, we have to explain how procedural meta-cognition based on procedural memory can be the basis for a kind of consciousness, since both are generally viewed as being unconscious. Given the diversity of unconscious processes, there is certainly room for different kinds to vary in just how unconscious they are. In fact, Mangan proposed that fringe states fall between truly unconscious information and content-rich focal (explicit) consciousness, and create a bridge between them.

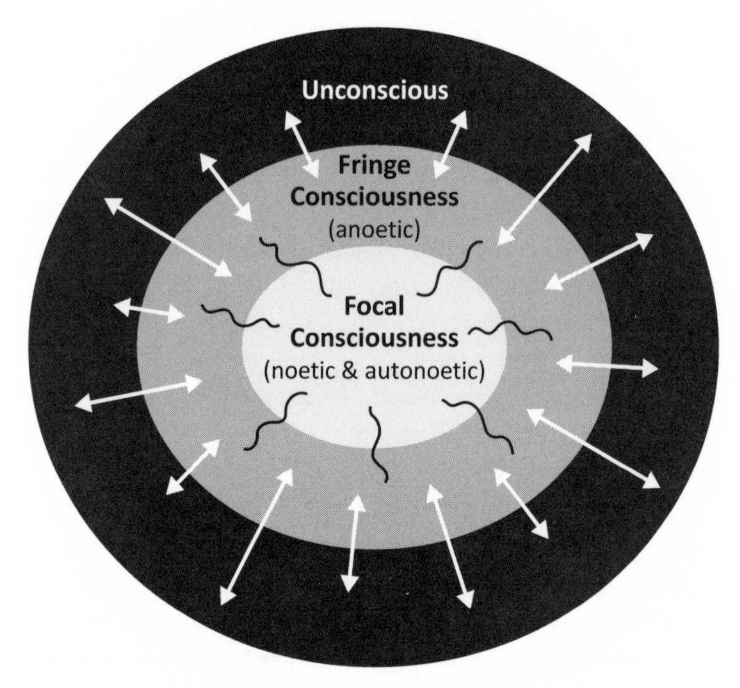

Figure 24.1. The relationship between the unconscious, fringe conscious, and focal-conscious states

It's a small leap from James's idea that fringe states allow explicit, content-filled conscious states to feel "right" to Tulving's idea that anoetic states accompany and color noetic and autonoetic states. Putting all this together, I suggest that fringe states endow noesis and autonoesis with feelings of familiarity by acquaintance, a feeling of rightness (or wrongness), and that this constitutes much if not all of what anoetic consciousness is, and does.

I would not have reached this conclusion without having noticed Vandekerckhove and Panksepp's connection of James's fringe with Tulving's anoesis. But I differ from them in an important way. For example, they treat the lower-order states of the amygdala as the source of fringe feelings of fear. But for me, the lower-order amygdala state has to be re-represented / re-described via the sub-granular meso-cortical PFC in order to enter the granular PFC mental model and make noetic awareness of danger feel like danger, and autonoetic awareness of noetic danger feel like fear.

Phenomenologists such as Edmund Husserl and Jean-Paul Sartre held that pre-reflective states enable reflective self-awareness. Building on Husserl and Sartre, Shaun Gallagher and Dan Zahavi proposed that without pre-reflective states there would be nothing that it is like to be reflectively conscious. If we allow that anoesis is a kind of pre-reflective state, we can rephrase Gallagher and Zahavi in this way: without anoetic consciousness, there would be nothing that it is like to be noetically or autonoetically conscious.

Anoesis and Affect

The term *affect* is often used in the study of emotion to refer to valenced (positive and negative) feelings that can vary in terms of their degree of arousal. For example, fear is a highly aroused negatively valenced emotional state, while sadness is a lowly aroused negatively valenced state. Affect may, it seems, be subsumed within the anoetic framework of states that feel right or not.

Anoesis and the "Ownership" of Mental and Body States

Timothy Lane has written about a patient who had conscious awareness of visual stimuli, but he lacked a sense of ownership of these perceptions. In terms of what we have been discussing, we might say that the patient experienced noetic consciousness of visual stimuli, but these states lacked a feeling of rightness, without which he did not feel he owned his perceptions.

Similarly, Stanley Klein has described patients in whom mental or bodily states, after brain damage, no longer feel "right" (no longer feel like they belong to them). Such examples, he says, bring into stark relief the importance of the subtle, mostly unnoticed states that normally make explicit conscious states feel right.

Given the earlier discussion, it seems that in healthy people anoetic fringe states make possible the tacit knowledge that explicit, content-laden, mental states are one's own, without having to explicitly affirm this. As William James and others have pointed out, "the feeling of an absence is different from the absence of a feeling."

Other evidence comes from states of depersonalization, such as in psychosis, or when one is under the influence of psychedelic substances. For example, Kalina Christoff and colleagues note that one can feel like a bystander, watching the body responses and mental states of another person, when in such a state. Of particular interest, Christoff's team identified a functional disconnection in such states between the meso-cortical PFC component of the default mode network (involved in body and mental state ownership) and the medial frontal pole (involved in cognitive processes that underlie one's inner narration of who one is). Importantly, the effects on ownership preceded the cognitive effects. This is consistent with the possibility that tacit anoetic awareness of ownership typically adds feelings of rightness to momentary explicit noetic and autonoetic awareness. This dissociation highlights the importance of keeping separate the sub-granular meso-cortical areas and granular neocortical areas in the medial wall of the hemisphere.

It is important to distinguish between conditions where one's body does not feel right from situations where the rightness of ownership is lost. For example, if you are nauseous, have a headache, are dizzy, or otherwise feel sick, your body does not feel right. Similarly, if you are depressed, your body can feel tired and / or heavy. But these feelings of wrongness about the body come with content-dependent explicit awareness about one's body states. In other words, in these situations anoesis, noesis, and autonoesis are interacting properly. But when body and / or mental ownership is lost, Christoff and colleagues argue that there is a breakdown in the coupling of anoesis to explicit conscious states.

Christoff and collaborators suggested something else quite interesting—that the reason psychedelics and meditation / mindfulness practices help relieve anxiety and mental suffering is because they temporarily decouple anoesis from noesis and autonoesis, thereby reducing one's identification with one's own thoughts and beliefs, and making them easier to change.

In summary, the overall contribution of the PFC to autonoetic consciousness involves circuits that cognitively re-represent / redescribe episodic and semantic explicit knowledge, and integrate it with tacit, fringe information about who you are. Anoesis, in short, can be thought of as a kind of primitive-mental-state consciousness that subtly colors explicit consciousness in barely perceptible hues of non-knowing experience. It is intermediate between the explicit, content-rich conscious awareness associated with autonoetic and noetic consciousness, and the even more primitive, and non-mental, condition of being alive, awake, and responsive to sensory events (that is, creature consciousness).

Deep Learning

But how does the activity of tacit procedural circuits, when metacognitively re-represented / re-described, provide the unconscious context that allows you to know what it anoetically feels like to be you? Deep learning may be a key part of the answer.

In the world of machine learning and artificial intelligence, the brain is thought of as a deep learning computer, or a collection of deep learning neural computers that use statistical learning processes based on prediction error-based reinforcement learning—learning from errors. These processes continuously and unconsciously update neural states when they deviate from past probabilistic regularity. Judea Peral, an AI pioneer, characterizes deep learning as the way our brains use past learning to make predictions that reduce uncertainty. Uncertainty, he says, prevails in even the simplest of decisions in life, such as crossing the street or speaking with other people.

The predictions made by these automatic learning processes help us cope without having to constantly think about what to do each and every moment.

Piotr Winkielman and his colleagues proposed that vague feelings and judgments, when re-represented, impact cognitive states and contribute to the coherence and fluency of the mind, and hence to the unity of consciousness. If these cognitive states refer to noetic and autonoetic states of consciousness, we would have an account of how meta-cognitive representation / re-description of deep-procedural learning provides the unconscious context that underlies fringe anoetic feelings. Anoesis, in other words, may, as implied earlier, be the elixir that makes your noetic and autonoetic experiences feel right—that makes red feel like red, and fear feel like fear.

But what exactly is being re-represented / re-described in anoesis? Hakwan Lau and I proposed that for modality-specific noetic perceptual experiences, the re-representation / re-descriptions underlying anoesis typically involve the activity of the sensory circuits that are presently engaged in perception. Because deeply learned activity of a certain type has routinely occurred in some set of these sensory circuits during similar situations in the past, the kind of sensory activity present now helps make the present experience feel familiar. But in our view the sensory cortex alone is not sufficient for anoetic consciousness. Also required is that the lower-order sensory information be re-represented / re-described via procedural meta-cognition, which brings anoesis into the higher-order PFC mental model.

More recently, Lau and I, in collaboration with Matthias Michel and Steve Fleming, elaborated on the idea that procedural learning by circuits within and between the sensory cortex and the PFC underlie the subjective experience of sameness (familiarity). Put in the context of the present discussion, anoetic states give explicit conscious experience a feeling of rightness via tacit comparison to other related experiences in a *higher-order subjective quality space,* but only when re-represented / re-described via implicit meta-cognition.

For example, over the course of your life, through many experiences with the color red, you have deeply learned what it feels like to see red, even though many shades qualify as red. First-order theorists such as Rafi Malach argue that you experience red when the wavelengths of what you're looking at match deeply learned wavelengths in the visual cortex that you have come to refer to as "red." Higher-order theorists, by contrast, say that these first-order procedural states must be meta-cognitively re-represented / re-described procedurally for you to experience what you know as red. Therefore, in higher-order theory, the noetic experience of an apple feels right, regardless of whether it is of a light or a very intense shade of red, if the present hues have been part of the experiences that taught your visual and prefrontal cortex, via recurrent connections between them, what red is to you.

Obviously, visual features other than color must also match past learning in order for the noetic experience of what seems to be an apple to feel right. If you have deeply learned the general shape of apples by encounters with many apples, a close approximation of the shape feels right, but a shape that significantly deviates from the deeply learned pattern (more cube-like than spherical) might feel discordant. Alex Cleeremans's radical plasticity higher-order consciousness hypothesis proposes something similar.

Anoesis is just there. Lacking explicit content, this kind of tacit experience, as Mangan pointed out, normally makes minimal demands on working memory and articulation. But when deeply learned predictions of what noetic states tacitly feel like are violated, we experience dissonance and turn to explicit, deliberative cognition to make things right. For example, an odd occurrence, such as the noetic perceptual experience of a purple apple, would not feel quite right, and may well require some explicit cognitive reconciliation, such as the thought that despite being purple, the object otherwise seems to be an apple.

From Anoesis to Autonoesis in Fear and Other Emotions

Emotions are the mental center of gravity of the human brain, fodder for narratives and folktales, and the basis of culture, religion, art, literature and relations with others and our world —of all that matters in life as we know it. A theory of consciousness that cannot account for emotions is hardly a theory of consciousness. I will explain how emotions emerge in my multi-state hierarchal higher order theory as autonoetic states of consciousness.

Circuits involving the amygdala, hypothalamus, and periaqueductal gray area together are often considered a mammalian *fear circuit*. By contrast, I argue that these constitute a defensive survival circuit that has ancient roots in vertebrate history. The defensive survival circuit controls behavioral and physiological responses that co-occur with fear, and feedback from these responses serve as signals that contribute to the feeling of fear. But the defensive survival circuit is not itself responsible for the feeling.

The feeling of fear, in my view, is an autonoetic experience that results from the cognitive interpretation that you are in a situation of danger based on the presence of a stimulus that you have, from semantic or episodic experience, come to know of as dangerous. The interpretation is conceptualized by momentarily active schema, including schema about danger and fear (emotion schema), but also schema about you and your relation to danger and fear (self-schema), all filtered through your culture's understanding of fear and danger (cultural schema). The core of the experience results from the meta-cognitive representation / description of the active schema in the granular PFC working-memory mental model. The conceptualization that danger is present biases the mental model toward the episodic conceptualization that you are in harm's way, and hence toward the pattern-completion of the realization that you are feeling fear—that you are afraid. This is a personally constructed, culturally constrained, autonoetic experience.

Emotion words categorize emotional experiences and provide conceptual anchors that help us understand and remember our experiences. These labels are not required in order to feel emotionally aroused, but they are required to feel the emotion named by the label. A distressed young child, lacking specific emotion words, cannot *not* experience herself as being in a state that an older child experiences as fear when her mental model, drawing on her emotion and "self" schemas (schemas about who one is), conceptualizes the state that way. But even in adults, the non-conscious underpinnings of emotions are not always precise enough to produce an experience that is clearly identified with a common emotion word. One may feel uncomfortable, concerned, or distressed in a situation, and not progress to something more specific. But as the situation unfolds and more information is collected, it is also possible that a vague feeling may turn into one labeled and experienced as fear, which might, with additional information, morph into anger, jealousy, or relief.

Earlier I mentioned Marie Vandekerckhove and Jaak Panksepp's alternative view of emotions. We agree that for explicit autonoetic fear in humans, anoetic experiences are in the background, overshadowed by the more prominent cognitive-based experiences. The main difference in our views is that for me, cortical re-representation / re-description by sub-granular, meso-cortical PFC is required for anoetic fringe feelings of fear. By contrast, for them, anoetic fringe feelings of fear arise directly from subcortical defensive circuits involving the amygdala and the periaqueductal gray area.

A wider range of lower-order procedural states and brain circuits are often re-represented / re-described via procedural meta-cognition in meso-cortical PFC circuits to make emotional experiences feel the way they feel than is the case for the experience of simple sensory perceptions. Remember that each kind of circuit undergoes deep learning of statical normality, and many more circuits are involved in emotional than perceptual processing.

Body states are often discussed as a key factor underlying emotional feelings. But it is important to note that not all instances of fear or other emotions come with predictable body responses. People can feel fear despite the absence of robust physiological arousal. When a feeling of fear is not accompanied by the body responses that our culturally and personally shaped schema tell us are supposed to occur, we can make the state feel right by a top-down, mental model-based cognitive fiat—"I must have been so afraid I didn't notice my heart racing."

(Continued)

From Anoesis to Autonoesis in Fear and Other Emotions *(continued)*

Underlying this may be what Antonio Damasio refers to as "as-if loops" that mentally simulate body feedback when it is absent.

I am totally on board with Damasio's "as-if loops" idea, but I go further. For me, all emotions are mental simulations—that is, all are psychological inventions (that is, narrations) based on mental models. Body feedback can be part of the lower-order amalgam, but it is not necessary. I do not subscribe to the distinction between innate basic and culturally acquired secondary emotions. Similarly, Lisa Barrett, Kristen Lindquist, Gerald Clore, Andrew Ortony, and other cognitive theorists reject this distinction.

Early in cognitive science, cognition and emotion were treated as opposites—cognition was thought to be based on conscious reason (it was rational), and emotion was based on unconscious passion (it was irrational). But cognition and emotion are no longer so narrowly conceived—cognition now includes non-rational processes such as intuitions and motivations. In addition, as just noted, even emotions are viewed by some, including me, as cognitive interpretations. A critic from the rational camp could argue that if fear relies on a rational cognitive interpretation, then why is one afraid when they know they have no objective reason to be so? But, as just noted, mental models are dynamic, and people can hold conflicting thoughts and beliefs unconsciously, and shift between them moment to moment in consciousness. Love can turn into jealousy, jealousy to anger, and anger to fear, all in a matter of seconds.

What good is consciousness, including conscious emotions like fear? Nathanial Daw and I have suggested that consciousness, including conscious fear, opens novel avenues of decision-making via conscious deliberation. This advantage, which depends on working memory, is admittedly negated when cognitive resources are drained by overwhelming danger in the moment, or by anxiety in the absence of actual danger. But just because conscious fear sometimes fails to help us does not mean that it never contributes to our goal-directed decision-making and behavior. Chris and Uta Frith, and their colleagues, have suggested both that deliberate social interactions require reflective self-awareness, and that autonoetic consciousness evolved in connection with human social interactions.

Are Anoetic States the Physical Manifestations of Qualia?

Given that anoesis is characterized by "ineffable," "vague," and "diaphanous" feelings that co-occur with explicit content-rich conscious states, it is tempting to suggest that anoetic states may offer a physical account of why it feels like something to be conscious of explicit conscious states. In other words, anoesis might constitute a physical account of qualia. I resist the temptation to apply the term qualia to anoesis, however, because the non-physical, dualistic, other-worldly features of qualia would inevitably add unnecessary conceptual baggage.

For example, the philosopher Michael Tye has recently published a book that touts a dual organization of consciousness that bears some similarity to Mangan's ideas. He postulates that a vague, content-lacking *what-it's-likeness* is transferred to an articulable content-rich kind of consciousness that we use in action and thought. While very much in keeping with the thrust of the discussion in this chapter, Tye, who is unable to see a way for the vague content-lacking what-its-likeness component to be physical, turns to panpsychism to account for it. But this philosophical hail Mary is unnecessary if anoetic feelings and the tacit phenomenal feel of explicit conscious experiences are one and the same.

25

What Consciousness Might Be Like in Other Animals

Human bodies differ in unmistakable ways from the bodies of other animals, including the bodies of our closest primate relatives. Our brains also differ in significant ways from the brains of other primates, and those primates' brains differ from the brains of other mammals. Given these variations, it should not be controversial to suggest that the mental states of other animals might also differ from ours. But boy is it controversial. In this chapter, I will discuss why, and suggest a scientific approach that may help us get past some of the points that fuel the controversy.

The Methodological Stumbling Block

The problem is not necessarily what our animal friends lack, but instead what we possess. Only humans have brains in which cognitive systems have been rewired by language, allowing us to both think and communicate in verbal terms. According to Descartes, humans demonstrate that they possess a rational soul (consciousness) through speech. The philosopher Daniel Dennett similarly noted that the hallmark of conscious states in humans is that we can talk about them. But people can also report on their conscious states nonverbally; for example, if you are asked which of three objects on a screen is an apple, you can say "the red one" or you can point to the apple. In the absence of brain pathology, most—if not all—of the time when a non-verbal report of a conscious experience can be given, a verbal report can also be offered. Although you may not be able to verbally describe with complete accuracy the entirety of a conscious experience, you can usually say *something* about it.

While we humans can typically respond both verbally and non-verbally to something we are conscious of, we can respond only non-verbally to unconscious information in our brain. As a consequence, the contrast between verbal and non-verbal responses is very useful for separating conscious from non-conscious behavioral control. Specifically, because non-verbal responses can be controlled either consciously or unconsciously in adult humans, they are ambiguous as readouts of consciousness (even when behaviors have symbolic value, such as a shrug, scorn, or a smile), whereas verbal reports are considerably more reliable.

Where does that leave pre-verbal children? The esteemed child psychology researcher Michael Lewis has noted that self-reflective consciousness begins to emerge between eighteen and twenty-four months. He points out that based on their behavior, children look as though they are consciously afraid long before they can have fearful or other emotional experiences.

The bottom line, then, from the scientific point of view, is that non-verbal responses are often inconclusive as evidence for conscious control. The most straightforward and reliable way to distinguish mental state consciousness from non-conscious processes that control behavior is, therefore, via verbal report. This is why verbal reports (barring forgetfulness, deception, or mental dysfunction) are considered the gold standard in assessing consciousness in humans, even if they are not perfect. And just as the presence of a verbal report is the best evidence that someone was conscious of something, the absence of the ability to give a verbal report about a stimulus is the best evidence that the person was not conscious of the stimulus. A shortcoming is that the veracity of verbal reports weakens as the time between the experience and the report increases, since it becomes more dependent on long-term memory than on the content of the working memory mental model.

Because of these methodological issues, researchers seeking evidence for consciousness in animals often turn to other, less satisfactory strategies. A particularly popular one is *argument by analogy with human behavior.* If an animal acts the way we do when we are conscious of some stimulus, it must be conscious as well. This approach has been criticized on the grounds that it tends to ignore leaner alternatives (such as associative learning) that explain behavior without recourse to consciousness. The critics do not offer non-conscious explanations because they want deny consciousness in non-human animals. Instead, they do this to encourage more rigorous standards for what one can and cannot say about consciousness from behavioral observations alone.

Echoes of Darwin's Scientific Anthropomorphism

The extensive reliance on analogy with human behavior as a means of scientifically demonstrating animal consciousness largely stems from Darwin and the social climate in which he lived.

Christoph Marty, writing in *Scientific American* in 2009, summarizes an 1838 exchange of letters between Darwin and his fiancée, Emma Wedgwood. Darwin told Wedgwood that his work was leading him to the conclusion that all of life descended from a common ancestor—that the biblical story of creation was wrong. She replied, "I implore you to read the parting words of our Savior to his apostles, beginning at the end of the 13th chapter of the Gospel according to John." Although they married, Wedgwood feared they would be separated in death—implying that she would be in heaven while he burned in hell.

Wedgwood's concerns had an effect. Darwin realized that treating human bodies as the product of biological forces, rather than as a gift from God, would be an uphill battle in a Victorian culture shaped by the Anglican Church. He became less vocal about his theory, and held back on publishing it until he could make a stronger case.

In 1859, more than twenty years after the exchange with Wedgwood, he published *On the Origin of Species by Means of Natural Selection; or, The Preservation of Favored Races in the Struggle for Life.* In it, he focused on the gradual evolution of animal bodies, hardly mentioning humans. Still, it did not go over well with the English people.

Darwin's next book, *The Descent of Man, and Selection in Relation to Race,* was not published for another decade. Finally he took the step, asking "whether man, like every other species, is descended from some pre-existing form." Using anatomical and mental similarities, he concluded that we had indeed evolved from animals. Most important for our purposes, he noted that "there is no fundamental difference between man and the higher mammals in their mental faculties."

But rather than cataloguing the animal features of human minds, he began to discuss animal minds in human terms. Why did he take this strange turn? Elizabeth Knoll, writing about Darwin's struggles, argued that he was troubled by the tepid, and sometimes hostile, reception to his views on evolution, and he hoped that this "more cheerful view" might help the public come to accept his theory.

This new approach was especially evident the next year in *The Expression of the Emotions in Man and the Animals.* In it, Darwin applied an anecdotal, commonsense (folk psychological) approach to the evolution of emotions. Given that we are afraid when we are escaping from harm, an animal fleeing from harm must feel fear as well. He also wrote about scorn, jealousy, pride, and contempt in dogs and other mammals. The psychologist Fred Keller has pointed out that Darwin's use of anecdotes to explain animal behavior in terms of human mental states was a glaring deviation from the "self-critical zeal that marked his biological endeavors."

Darwin's anecdotal approach had a tremendous influence on the budding field of animal psychology in the late nineteenth century. Georges Romanes, a leading figure and devoted follower of Darwin, announced that behavior is an ambassador of the animal's mind. Rampant use of anecdotes about the mental causes of animal behaviors continued into the early twentieth century and was one of the leading factors that led to the rise of behaviorism, and hence the rejection, for decades, of mental states as explanations of behavior in psychology.

As the influence of the behaviorists began to weaken in the mid-twentieth century, anthropomorphic explanations of animal behavior unfortunately returned. Rats were no longer pressing bars for reinforcers, but for "pleasure," and when in danger they froze out of "fear." This trend continues today: terms humans invented to describe our kinds of mental states are widely, and often indiscriminately, used to explain animal behavior in scientific contexts. Anthropomorphism is, in fact, proudly embraced by some scientists.

When claims about animal consciousness are based on intuitions and beliefs that match commonsense and lore, they feel correct. And when they are repeated authoritatively by scientists, they come to be treated as so obviously factual that no reasonable person could possibly question them.

Scientists like me who propose interpretative restraint have been mocked as "deniers" of animal consciousness. But so-called deniers do not in fact typically argue that animals lack conscious experiences. Suggesting that behavioral data held up as "obvious proof" of conscious mental states is inconclusive does not make us deniers.

Contemporary anthropomorphism thrives in no small part because the father of modern biology gave credence to the idea that it is scientifically permissible to assign humanlike emotions and other mental states to animals based solely on similarities between their behaviors and ours. But as others have pointed out, being a great biologist did not make Darwin a great psychologist. The science of emotion, in fact, may have ended up being very different had he stayed in his lane. Yet if he had not used anthropomorphism to his advantage, his theory of evolution may not have succeeded—and life today might be quite different had biology, and hence medicine, not had the benefits that Darwin's revolutionary insights made possible.

Why Are We So Anthropomorphic?

In *The New Anthropomorphism,* J. S. Kennedy offers an account of why we are so prone to anthropomorphism:

> Anthropomorphic thinking . . . is built into us. . . . It is dinned into us culturally from earliest childhood. It has presumably also been "pre-programmed" into our hereditary make-up by natural selection, perhaps because it proved to be useful for predicting and controlling the behavior of animals.

Our language is strongly anthropomorphic, he said, and, as a result, our concepts and thoughts tend to lean in this direction as well. If Kennedy is right, anthropomorphism is part of human nature, and is perhaps why we all see humanlike emotions in our pets. And that's okay. Scientists don't have to wear their scientific hat every moment. But they must put it on when being a scientist.

Often, scientific arguments for animal consciousness rest on intuitions and beliefs. The philosopher Bertrand Russell once noted, "All the animals that have been carefully observed have behaved so as to confirm the philosophy in which the observer believed before his observations began."

But just because an animal behaves in a way that a human might in response to danger, or to other biologically significant stimuli (food, a sexual partner), does not necessarily mean that it is having a conscious experience similar to what a human has in a similar situation. The kinds of behaviors said to be due to conscious states in other animals can, in fact, often be accounted for by cognitively or behaviorally leaner explanations (explanations that do not involve conscious control of the behavior).

We all assume that a dog writhing and yowling on the side of the road after being hit by a car is horribly distressed. And it surely is. But the behaviors we see are reflexive responses and do not, themselves, reveal the distraught mental state of the animal.

Because we are fundamentally anthropomorphic, we often can't help calling on our understanding of how our own minds work to understand what infants or animals might be consciously experiencing. Assuming that they are feeling something akin to what we might feel in a similar situation is the appropriate moral reaction. But it is not an appropriate scientific response. Because reflexive behaviors are mere neurobiological-realm responses, they cannot be used as scientific measures of conscious experience. Neither are mere cognitive-realm behaviors conclusive about consciousness, since they can be controlled consciously or non-consciously. To show that a given behavior is controlled consciously, consciousness itself must be measured, and non-conscious alternatives must be ruled out.

The philosopher Brian Key, writing specifically about fish, has made a very similar point about why we can't use behavioral responses as a definitive measure of consciously experienced pain in animals. He's not saying that fish don't feel pain. He is instead arguing that the anthropomorphic *naïve intuitions* we are all prone to when observing behavior responses to aversive stimuli are not enough to conclude that the animal is subjectively experiencing pain.

You may be wondering, why are we on firmer ground studying consciousness in humans? Barring a brain disorder, all humans are born with brains that possess the same basic structural components and functional capacities—all the same realms of existence. Given that, if I have the capacity for inner awareness of my mental states, I can with some confidence assume that you have this capacity as well.

Mike Gazzaniga has described consciousness as an instinct. This consciousness trait, like any species characteristic, will vary across individuals, but will be present, or at least is potentially present, in all members of the species. Because no other animal has our kind of brain, scientists should be cautious in generalizing from our experiences to theirs.

How to Minimize the Methodological Barrier

When I am being a regular person, rather than a picky scientist, I am perfectly comfortable talking about animals in anthropomorphic ways. And even when being scientist I am comfortable saying that that some animals might, or even likely do, have conscious experiences. But because of the methodological barriers that prevent clear proof, I use modifiers such as "might" and "likely," since it is not possible to conclude with certainty whether they are conscious, and if they are, what it is like to be them.

It may not be possible to fully eliminate the methodological barriers. Nevertheless, there are probably ways to minimize them. The philosopher Matthias Michel recently proposed that "the study of non-human animal consciousness should start . . . in humans by identifying those capacities for which conscious mental states make a difference in virtue of being conscious mental states."

Jonathan Birch and colleagues have made several efforts along these lines, most recently by suggesting that features of human consciousness—such as perceptual richness, integration across time, and self-awareness—be looked for in animals. "Self-awareness" is the only one of these features that is about consciousness itself, and it has proven to be notoriously controversial as an indicator of consciousness in non-human animals. The other features are mere behavioral correlates of consciousness in humans. It would certainly be useful to assess these correlates in different animals, so long as the results are not used as hard evidence for consciousness.

I have been working on a different empirically based approach for understanding what mental state consciousness *might* be like in other mammals. I start by considering the brain mechanisms of anoetic, noetic, and autonoetic consciousness in humans. This then provides a way to reverse engineer what kind or kinds of consciousness might be present in different mammals, given anatomical similarities and differences between their brain and ours.

For example, sub-granular meso-cortical PFC areas are shared by all mammals. Therefore, whatever kind of consciousness these areas enable in humans may also be present, at least to some extent, in non-human primates and in non-primate mammals. Since all anthropoid primates (monkeys, apes, and humans) have granular areas of PFC, but other mammals do not, whatever roles that granular PFC areas play in human consciousness may also be present in some form in other primates, but would be lacking in non-primate mammals. Because early evolving primates (that is, prosimians and tarsiers) have only the primitive beginnings of granular PFC, their conscious capacities, if they have them, likely fall between what monkeys and non-primate mammals possess. Finally, contributions of human-unique areas of granular PFC, especially within the frontal pole, would be lacking in other primates (except possibly some other great apes) and other mammals. Figure 25.1 shows the location in

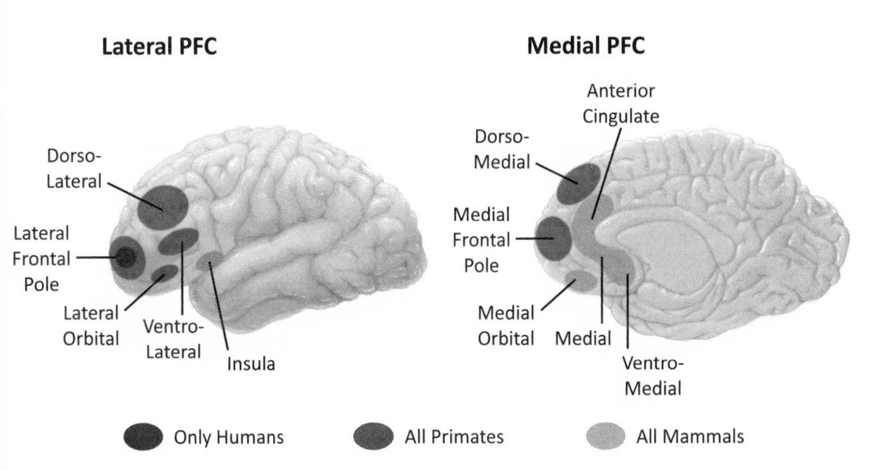

Figure 25.1. Prefrontal cortex areas possessed only by humans, only by primates, and by all mammals. From Joseph E. LeDoux, "What Emotions Might Be Like in Other Animals," *Current Biology* 31 (2021): R821–R837, figure 3.

the PFC of the common mammalian areas, areas particular to primates, and an area unique to humans.

I propose that areas of sub-granular meso-cortical PFC contribute to anoetic fringe conscious feelings in all mammals; areas of lateral and medial granular cortex contribute to noetic factual and conceptual consciousness in anthropoid primates; and the lateral frontal pole contributes to autonoetic consciousness and its capacity for mental time travel in humans and possibly some other great apes.

The virtue of this approach is that the hypotheses are tied to empirical data about the relation of mind to brain, rather than on presuppositions about what we think the relation of behavior to mind should be in other animals. For example, if the human-unique component of the lateral frontal pole indeed turns out to be essential for a human to know that it is the entity having an autonoetic experience, this would explain why it is so difficult to demonstrate robust self-awareness in other animals, including other primates.

Similarly, if lateral PFC areas, such as the dorsolateral region, are important for noetic factual and conceptual consciousness, we would have clues to how consciousness in monkeys is similar to human consciousness and different from consciousness in non-primate mammals. Finally, if meso-cortical PFC makes anoetic consciousness possible in humans, we would have insights into shared fringe anoetic experiences across all mammals.

Several caveats are in order. First, I am not suggesting that consciousness lives in these areas. Instead, what I am saying is that circuits in these areas make important contributions to consciousness by virtue of their various connections with each other, and with other areas. Second, as discussed in Part IV, the primate brain continued to evolve as primates diverged from other mammals, and the human brain continued to evolve as humans diverged from other primates. Therefore, that animals possess similar areas does not mean that they have identical areas and functions—anoesis may be similar across all mammals but somewhat different in primates from other mammals, and different in humans than in other primates. Third, and related to the previous point, the addition of language to the human brain was a novel event—a revolution, rather than an instance of continued evolution—and it greatly affected our capacity to think, plan, decide, understand, and feel. Fourth, even if all my anatomical hypotheses prove to be too simplistic, or even wrong, the basic strategy outlined here can be adapted to pursue other anatomical hypotheses in relation to anoetic, noetic, and autonoetic consciousness, or in relation to other ways of classifying conscious mental states.

Consciousness with and without Warm Blood

My view of mental-state consciousness is tied to the kinds of cognitive re-representation / re-description processes that seem to be limited to true endothermic animals, including mammals and birds (see

Chapter 17). But just as endothermy does not guarantee model-based cognition (only some birds possess it), neither does having the cognitive underpinnings of consciousness guarantee consciousness. In mammals we can at least use homologies in brain organization in order to extrapolate from circuits underlying mental modeling to speculate about consciousness. We have to be more conservative in drawing conclusions about humanlike consciousness in birds because mental modeling has only been found in some birds, and therefore it likely reflects parallel, so-called convergent, evolution due to a combination of warm-bloodedness and species-specific selective pressures.

Given that I am trying to understand human mental-state consciousness, focusing on the relation between humans and other mammals is roughly where I want to be, evolutionarily speaking. I see the point of considering birds, reptiles, amphibians, fish, since our brains are related to these, directly or indirectly, in well-understood ways. And I have no problem with others speculating about consciousness in various protostomes (for example, bees, flies, and octopuses), or even in protozoa and bacteria. But that's not how I want to spend my time. That said, I feel I should devote a few words here to the topic of sentience, which often comes up in discussions of animal consciousness.

Sentience and Science

The term sentience is often used to refer to the conscious experience of external sensory information and, for some philosophers and researchers, includes sensations from interoceptive signals. To the extent that cognitive re-representation or re-description is not generally deemed necessary, sentience could, in principle, exist in all animals, regardless of whether they are warm- or cold-blooded, and regardless of whether they have robust cognitive capacities. Indeed, sentience has been proposed to be a pan-animal, and in some cases, a pan-organism, condition. But lack of clarity about what "sentience" actually means makes the claims difficult to evaluate.

The simplest definition of sentience is the sheer ability to respond to sensory stimuli. As I noted earlier, a paper by Merker is often cited as evidence for sentience in animals. But in using responsiveness to sensory stimuli as the criterion, Merker only showed instances of creature consciousness (the condition of being alive and responsive to sensory stimuli). What most who use the term sentience are trying to get at is the possibility that sensory-motor responsiveness demonstrates that animals have the phenomenal feeling of what it is like to be conscious, something we humans assume, from our experiences, that we possess. But mental state consciousness cannot be demonstrated by measuring sensory responsiveness. Others point to the ability of animals to engage in complex activities such as problem solving to make the case. Although this goes further behaviorally, problem-solving abilities, even when quite impressive, are alone not evidence for the existence of phenomenal feelings.

In the previous chapter I argued that anoesis is a primitive form of mental-state (cognitive) consciousness that results from re-representation / re-description in the meso-cortex of lower-order information about sensory, memory, and / or body states. I also suggested that anoesis is what makes noetic and autonoetic conscious states feel like something in us. If that's what anoesis does, maybe we don't need sentience to explain what it feels like for *us* to be conscious.

But aneosis so conceived may only exist in animals that have at least mesocortical procedural meta-cognitive (re-representation / re-description) -based anoetic consciousness, which may limit such states to mammals. Yet, if correct, it might pave the way for asking whether there might be a primitive—perhaps first-order—kind of state (sentience) that in humans and other mammals is overshadowed by anoesis the way that anoesis is overshadowed by noetic and autonoetic consciousness in us. Such states may exist between anoesis and creature consciousness, which Marie Vandekerckhove has also proposed in relation to a continuum of consciousness.

While anoesis has limited, tacit mental content (e.g., rightness, familiarity, ownership, confidence), sentience, so conceived as a first-order (non-cognitive) neurobiological state, would have even less content, and this may vary to different degrees among animals within and across vertebrate and invertebrate lineages.

Being a non-cognitive state might seem to eliminate the possibility of sentience having any mental qualities. But first-order theories of human consciousness (see Chapter 21) are non-cognitive mental state theories. While I believe that a higher-order explanation is required for complex human experiences, it is possible that autonoesis and noesis not only overshadow anoesis, but also obscure primitive underlying non-cognitive mental states in us. I don't subscribe to this idea, but neither do I completely rule out this possibility. If sentience and first-order mental states turn out to be one and the same in humans, it would open the door for first-order sentience having a long evolutionary history. But unless first-order sentience can be shown to exist in humans by differentiating it from both anoesis and creature consciousness, it will remain a vague notion that lacks clear scientific grounding as a kind of primitive conscious state.

My intent in discussing sentience was not to suggest a solution to what it is, but to illustrate where it might fit within the conceptual framework I have built in this book. I believe that if speculations about sentience and other forms of consciousness in non-human animals were more openly acknowledged as speculations, rather than treated explicitly or implicitly as facts that are so obvious, so intuitively correct, that they must be true, the work would be more widely respected and impactful. As Jonathan Birch, a leading figure in the field of sentience, recently put it, "animal advocates sometimes retreat into unjustified certainty. If you're certain, you can block out the gnawing thought that you might be wrong."

26

The Stories We Tell Ourselves, and Others

I am now at the part of the book where I explain how I think human conscious experiences (our perceptions, memories and beliefs, joys and sorrows, hopes and dreams, fears and fantasies) actually come about. The explanation I will offer has, in some sense, been in the works since my days researching split-brain patients in the 1970s. Back then, as I recounted in the Introduction, Mike Gazzaniga and I proposed that we humans construct interpretations that help us make sense of who we are, what we are doing, and why we are doing it. These interpretations, we suggested, are necessary to maintain mental cohesion, a sense of conscious unity, despite the multiplicity of brain systems that control our behavior non-consciously.

Over the subsequent decades, Mike continued to explore this idea through studies of split-brain patients, and came to refer to the underlying mechanism as "the interpreter." For my part, I turned to studies of rats to learn more about the non-conscious brain systems that might compel such interpretations in emotional, especially fearful, situations. I closely followed developments in cognitive science about working memory, and in philosophy about consciousness, and built on these to better understand how cognitive interpretations might come about in relation to human consciousness. My current version of the interpreter hypothesis proposes that the non-conscious content of working memory mental models is translated into conscious experiences by inner narrations. In this chapter, I will build the case that these narratives do not come in a verbal or visual or any other kind of recognizable code. They take the form of a modality-independent, or "a-modal," neural code, something called a mentalese, that not only supplies our conscious content but also controls our speech and actions.

Narrating Life

In *The Story-Telling Animal,* the literary scholar Jonathan Gottschall writes that our stories make us human. Novelists, inspired by William James's description of the "stream of consciousness," have capitalized on the narrative quality of human mental life by telling stories in ways that mimic the way that our minds work. James's novelist brother, Henry, nicely illustrated everyday narrations by saying that "adventures happen to people who know how to tell it that way."

Because autobiographies are based on the author's episodic memories, they are an explicit form of self-narration. But because they often diverge from the actual experiences recounted, they famously blur the distinction between life and fiction. Earlier I mentioned that even vivid episodic memories are not necessarily accurate; they capture the gist more than the exquisite details. The writing category of creative nonfiction explicitly embraces a loose connection to reality in the telling of stories.

Mark Freeman, who writes about the psychology of narration, suggests that autobiographies and memoirs are often somewhat fanciful because of the literary character of the writing forms. That may well be true, but there are certainly other factors at work, especially the ability of the human mind to create mental models that allow us to imagine alternative narratives of the past, and to re-store, or re-consolidate, memories, allowing the latest version constructed to seem like the "real" version. The more times we re-imagine and re-consolidate, the more opportunities we have to reconstrue the past.

Although philosophical concepts abound in the moral and ethical inner struggles of characters in novels, it is a two-way street. Philosophical theories, all theories in fact, are stories, ideas, about how something works. But some philosophical theories give narration an explicit, explanatory role. For example, Daniel Dennett proposed that from temporary multiple drafts (narratives) a more enduring self-narrative emerges and underlies what people refer to as their conscious experiences. Owen Flanagan, writing with Gillian Einstein in *Narrative and Consciousness,* observed, "Human beings are only persons in so far as they can hold in their heads, and tell, the stories of their lives." And Alasdair MacIntyre noted that "we all live out narratives in our lives and understand our own lives in terms of the narratives that we live out."

Scientists, too, have been interested in the narrative quality of the mind. For example, Jerome Bruner, one of the early pioneers in cognitive science, emphasized that one's understanding of oneself is "storied," or narrative, in structure:

> When you ask people what they are really like, they tell a great many stories involving the usual elements of narrative . . . there is an agent engaged in action deploying certain instruments for achieving a goal in a particular scene, and somehow things have gone awry between these elements to produce trouble. The stories they tell, moreover, are genre-like: One encounters the hero tale, . . . the tale of the victim, the love story, and so on. If one ever doubted Oscar Wilde's claim that life imitates art, reading autobiographies lessens the doubt.

A person's sense of being a psychological entity that persists over the course of their life, according to Bruner, is achieved by perpetually rewriting their narrative to make it fit with present circumstances, then carrying the revisions forward though memory. Bruner claimed that these rewritings reduce the cognitive dissonance caused by discordant thoughts and feelings that challenge existing narratives:

> Could there be any human activity in which the drive to reduce cognitive dissonance is so great as in the domain of "telling about your life"? The stratagems employed for weaving webs across translocations and dislocations, the degree to which memory search is guided by these stratagems, the absence of discomfort about what must have been glaring discontinuities, the extent to which culturally familiar narrative forms promote this seeming unity in multiplicity—all these give remarkable testimony to Leon Festinger's (1957) powerful insight about dissonance reduction as a principal engine in cognitive functioning.

Bruner's text is reminiscent of my graduate work on split-brain patients mentioned earlier. Mike Gazzaniga and I conceptualized confabulations of split-brain patients as a means of preventing or reducing cognitive dissonance and maintaining a sense of psychological coherence in the face of behaviors that are incompatible with what they consciously know. But we extended this idea to all of us—that we interpret or narrate our lives to control dissonance. If there is a difference between *confabulations* and *narratives,* it is that confabulations are efforts to compensate for a neurological or psychological problem, whereas narratives are routine, quite normal explanatory processes we all use in life.

Linda Örulv and Lars-Christer Hydén suggest that healthy people have many routine narratives to call on, but people with dementia have far fewer, and turn to confabulations when memory fails. Asaf Gilboa, Morris Moscovitch, and their colleagues have done extensive research on confabulation in patients with amnesia. Others have studied confabulation in autism spectrum disorder and schizophrenia. The sleep researcher Allan Hobson has argued that dreaming is a potent, naturally occurring form of narrative / confabulation, not unlike hallucinations in schizophrenia.

According to Lisa Bortolotti, "People who confabulate know what their attitudes and choices are, but do not have access to key information about the formation of those." Örulv and Hydén identified several problem-solving functions of confabulations: sense-making to understand the present; self-making to maintain personal identity; and world-making to organize interactions with the outer world. One way to think about all this is that confabulations are called on when ordinary narrations are insufficient to sustain a Jamesian feeling of rightness.

Healthy people use episodic memory to construct narratives that connect elements of thoughts about events into a coherent understanding of who one is and why they do what they do. This is dramatically illustrated by reflections from a patient who, because of brain damage, temporarily lost episodic memory and the ability to create narrations, including confabulations: "I did not have the ability to think about the future—to worry, to anticipate or perceive it. . . . Thus, for the first four or five weeks after hospitalization I simply existed." A case with a more enduring condition was Clive Wearing, who had a severe, and fairly specific, loss of episodic memory. He could remember, and narrate, only those events that had occurred in the previous thirty seconds.

Absent inner narrations, our understanding of our own folk psychological mind, and its role in understanding the minds of others (theory of mind), would not exist. As Bruner noted in "Life as Narrative," folk psychology endows the expected and ordinary events in life with legitimacy and authority, and when these beliefs are violated or challenged, new narratives are constructed.

In *The Self Delusion,* Gregory Berns nicely captures the contributions of episodic memory to personal narratives:

> Everyone has at least three versions of themselves. The first, situated in the present, is the one you're most accustomed to thinking of as "you". . . . but the present is an illusion. Present-yous just don't last long. . . . Even when we think we're living in the present . . . we're stuck in the past. This is the second version of you. . . . Although future-you is by its nature fuzzy, its function is both pragmatic and aspirational. When we think we're in the present, our brains are not only processing events that have already happened but also forming predictions about the immediate future. . . . Normally past-, present-, and future-yous combine seamlessly into a unified existence. This, too, is an illusion. So, who are you? The answer is: whoever you think you are. . . . Our brains construct narratives for our lives and . . . this process constructs our self-identity.

Narrative Content

How is the content of our narratives generated? We know that memory is crucial. But while undeniably important, it is not sufficient since other processes intervene between the retrieval of memories and the construction of our inner narratives.

In his article "Death of the Author," the literary theorist Roland Barthes wrote that "it is language which speaks, not the author." Tracing the history of writing, he argued that authors and readers are part of a larger cultural story (narrative) that is subsumed within language, and that the author communicates with readers through this shared language and the cultural knowledge it contains.

I came across Barthes's quote out of context, and my thoughts went to a different place. I was prompted to think about what happens in our brains when we talk or write. Words typically come out as sentences without you choosing each individual word and its grammatical placement—sometimes you do, but that is the exception. Most of the cognitive work when talking or writing occurs sub-rosa, that is, by non-conscious processes. Re-phrasing Barthes, "It is often a non-conscious mental model which speaks, not the conscious person." In this sense, my conclusion is not so different from that of Barthes, since we both remove the conscious person from at least some aspects of speaking and writing.

In sum, you can obviously be conscious of what you say or write, but not because you consciously said or wrote it. Verbal expression and conscious experience are, in my scheme, separate, that is, parallel, outputs of a non-conscious mental model and its non-conscious narrative.

Narratives Are (or Can Be) Pre-Conscious and A-Modal

Several times in this journey I have mentioned that every conscious state is preceded by non-conscious processing. But it's time to be more precise about this. The non-conscious states that precede conscious experiences are better termed *pre-conscious,* since not all non-conscious states are antecedents of consciousness.

The natural way to think about inner narratives is in terms of language. For example, the psychologist Julian Jaynes, in his provocative late 1970s book *The Origin of Consciousness in the Breakdown of the Bicameral Mind,* proposed that auditory hallucinations of inner voices (narrations) were a precursor to the evolution of consciousness. Perhaps the voice in your head, the verbal monologue that runs through your mind, is the narration that underlies conscious experience. Support for this is the fact that we can attend to and be conscious of the monologue—in fact, sometimes we can't shut it up. The inner monologue, which is, in effect, linguistic mind-wandering, has a role in verbal consciousness. But that is only a small part of the story I am telling.

Early behaviorists viewed inner speech as subvocalized language that slowly becomes quieter as one ages and passes for what people refer to as their mind. This idea was challenged in the 1930s by the Russian psychologist Lev Vygotsky, who gave *private speech* a defining role in the mental development of children, and in the mental well-being of adults. Consistent with Vygotsky's point of view, some psychologists today think of private speech as making crucial contributions to the emergence in childhood of a subjective sense of who one is, and to the continued maintenance of this personal awareness over time.

Private speech has also been related to working memory and executive functions, as well as to theory of mind, and consciousness. Perhaps it is not surprising, then, that when private speech is lost as a result of brain damage, the sense of who one is suffers. Recall the patient described earlier who, while recovering from brain damage, was unable to think about the past or future, and therefore "simply existed."

Not everyone has a verbal narration running through their minds. In fact, some people are more visually oriented than verbal. Actually, humans are believed to have used pantomime and mimicry as visual communication tools before language evolved; these talents are even thought to have been the foundation for the evolution of language. While people differ in their visual and verbal abilities, they typically have both. We are visual creatures because we are primates, and verbal creatures because we are humans. In film, the visual narrative can be as important as, or even more important than, the spoken script. Remember, the first films told their stories silently with images.

One possibility, then, is that we have different generators for verbal and visual narrations and use them on an as-needed basis. The articulatory loop and visuo-spatial scratchpad in Baddeley's original working-memory model are obvious candidates for verbal and visual narrative generators. But they are hardly the only kinds of narratives our brains produce. Every sensory system is a potential narrative modality. These sensory-specific narrative processes, though, are not sufficient to account for more multimodal, Gestalt-like states in which the whole, rather than its parts, are experienced. This requires the binding of components within and between sensory modalities, and also the binding of these with memory into larger wholistic amalgams.

For example, within-modality features of visual objects (the color and shape of an apple) are bound together in the visual cortex. Then multimodal binding involving convergence zones in the parietal and temporal lobes integrates the visual representations with representations from other modalities, such as audition. Episodic multimodal visual scenes are constructed in parietal and hippocampal circuits. The final construction of the multimodal Gestalt representation requires the assembly of schema in the sub-granular mesocortex. These schema are then integrated into granular PFC working memory, providing the conceptual foundation for the pre-conscious mental model of the present moment.

The output of the pre-conscious model, I suggest, results in a conscious experience with multimodal content, much like Baddeley suggested for his hypothetical episodic buffer. Baddeley stopped short of explaining how the episodic buffer might enable complex conscious experience. I will pick up where he left off, focusing on how complex multimodal experiences are constructed as higher-order states.

My proposal, in brief, is that conscious experience is not a direct output of the pre-conscious mental model but is instead secondary to an abstract (that is, multimodal, or modality independent), pre-conscious, narrative code, a kind of *mentalese,* that the model generates. But before I explain the mentalese nature of the narrative, I need to explain mentalese itself.

Mentalese

As originally proposed by the philosopher Jerry Fodor, mentalese is a version of what philosophers traditionally refer to as a *language of thought.* Mentalese is similar to our natural language capacity in that it has words with semantic meaning that can be flexibly combined into complex elements (phrases or sentences) by syntactic rules. But unlike the verbal code of natural language, mentalese is an abstract, generic code that can both generate thoughts (concepts) and flexibly reassemble them into more complex thoughts. The abstract—that is, modality-independent nature—of these thoughts allows them to be used to cognitively represent (have thoughts about) combinations of all varieties of external and internal sensory stimuli, as well as thoughts about perceptions and memories, and even the thoughts themselves.

Misunderstanding the abstract generality of the mentalese narrative has resulted in mentalese being criticized on the grounds that it requires that people have cognitive access to mentalese words and their meanings. But, as Michael Rescorla has pointed out, "Thinking is not 'talking to oneself in mentalese.' A typical thinker does not represent, perceive, interpret, or reflect upon mentalese expressions. Mentalese serves as a medium within which her thought occurs, not an object of interpretation. We should not say that she 'understands' mentalese in the same way that she understands a natural language." This is the sense in which I use mentalese here.

Jake Quilty-Dunn and colleagues have recently noted that some view the language of thought as an antiquated idea from a simpler time. But they argue that the language of thought hypothesis has quietly become the "best game in town" when it comes to understanding mental representations.

Mentalese and the Brain

Fodor was a *functionalist.* This means that he viewed the relationship of the mind, including its mentalese, to the brain as comparable to the way that software relates to the hardware of a computer—that is, the software has its own rules of operation that process information independent of the hardware. Given this emphasis on cognition as information processing, he and many other early cognitivists were not particularly interested in consciousness. Fodor summarized his views about consciousness by titling a 1991 article "Too Hard for Our Kind of Mind?" In it he writes: " Cognitive scientists mostly think that consciousness is a damned nuisance. . . . Or they don't think about consciousness at all."

Although findings about the relation of mind to brain were irrelevant for Fodor, others have pursued the significance of mentalese in the brain, including in relation to consciousness. For example, the neuroscientist Edmund Rolls, who has done pioneering research on the meso-cortex in relation to emotion and goal-directed behavior, proposed a higher-order syntactic theory of consciousness. It is quite similar to a mentalese theory to the extent that it argues that syntactic manipulations of semantic representations underlie mental states. But Rolls's theory is based on the brain's natural language system, rather than on an abstract mentalese.

By contrast, the philosopher Susan Schneider has offered a language of thought (that is, a mentalese) hypothesis tied to the global workspace theory of consciousness. Her version of mentalese, in contrast to Fodor's, is therefore quite compatible with modern neuroscientific understandings of the mind and brain. A more refined view of the relation of mentalese to consciousness might result if Claire Sergent's global playground—in which we mind wander about external stimuli, body states, and mental states—were weaved into Schneider's hypothesis (see Chapter 21).

Steven Frankland and Joshua Greene have proposed another brain-based language of thought hypothesis that I am particularly fond of. In their hypothesis, abstract concepts are formed and combined by the default mode network. Rather than the default mode network itself, I prefer to emphasize some of the specific areas they implicated. Included are neocortical temporal and parietal lobe areas involved in semantic / conceptual and episodic memory; sub-granular (meso-cortical) PFC areas involved in memory and schema processing; and granular (frontal pole and dorsolateral) PFC areas involved in working memory and other aspects of higher cognition. They propose that these interconnected areas create representations using a Tolman-like cognitive map, rather than the sentence-like syntactic structures in Fodor's language of thought. Key to their hypothesis are grid cells, which were mentioned in Part IV in relation to spatial maps in the temporal lobe. What Frankland and Greene specifically argue is that grid cells in the temporal and parietal lobe, and their interactions with sub-granular and granular PFC areas, allow abstract, *a-modal* (that is, modality independent) neural coding to contribute to conceptual maps and mental simulations. Their hypothesis, which treats the a-modal concepts as higher-order consequences of integration across modalities, seems preferable to other hypotheses that emphasize modality-specific abstract concepts.

Frankland and Greene steered clear of consciousness. But because the circuitry they emphasize overlaps extensively with the circuitry of my multi-state hierarchical higher-order theory of consciousness (see Figure 22.1), it offers a possible neural account of the a-modal narrative output of the mental model and its role in integrated, complex conscious experiences.

The Mentalese Narrative Stream and Its Distributaries

The starting assumption in my mentalese theory is that a higher-order state is established in a pre-conscious, granular PFC, working-memory mental model (Figure 26.1). The content of the mental model reflects its momentarily active lower-order input state or states (sensory, mnemonic, schematic, linguistic, goal value, homeostatic, and others).

If the only thing occupying the pre-conscious mental model is a visual stimulus, or a memory, or linguistic mind-wandering (inner speech), then that is what determines the content of the narrative. If multiple distinct kinds of inputs are entering and being integrated

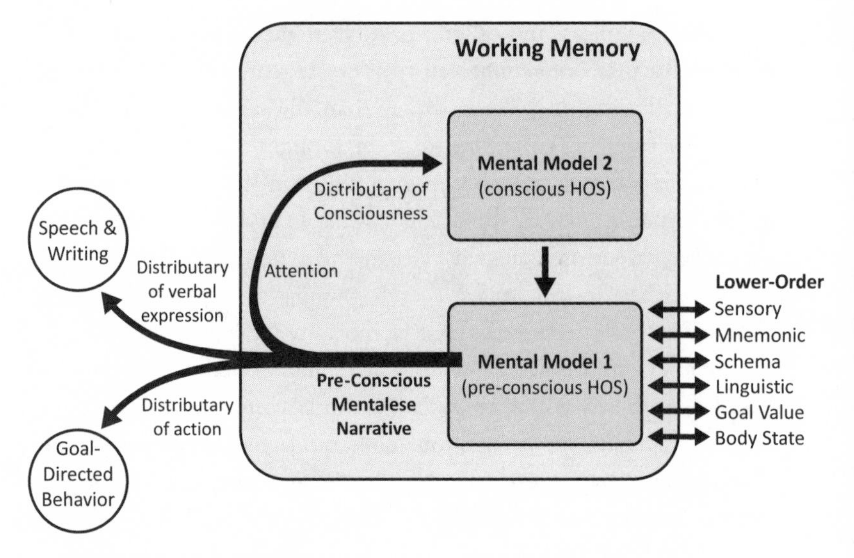

Figure 26.1. A mentalese narration separately underlies explicit mental-state consciousness, verbal reporting, and goal-directed behavior

by the mental model, as is often the case in real-life situations, then the content results in a narrative with correspondingly more complex content.

In-the-moment mental states require momentary neural activity consisting of short, coherent neural events that reflect what happens over the course of several hundred milliseconds. These are bundled together into *event chunks* by *postdictive processing,* as discussed previously. In other words, the pre-conscious content of the present-moment mental model (that is, the high-order state) is an event chunk.

One set of outputs of the pre-conscious mental model goes back to the lower-order input circuits, setting up recurrent activity that updates and sustains the mental model (bottom right two-way arrows). Key to my idea, though, is that the other output of the pre-conscious mental model is an abstract, a-modal, mentalese narrative about the present moment.

The modality-free or a-modal feature of the narrative allows its conceptual content to be used by diverse downstream processors. For example, it allows one to respond verbally through speech, writing, or sign language, or non-verbally through a wide variety of distinct goal-directed behavioral actions (walking, running, swimming, climbing, hitting, hugging, pointing, waving, sticking up one's middle finger, frowning, smiling, sneering, or chuckling). But the mentalese narrative also supplies the content of conscious experience.

The abstract mentalese narrative can, in other words, be thought of as a *mental stream* with three broad distributaries, or sub-streams, that diverge from the pre-conscious mental model (Figure 26.1). One is the *distributary of verbal expression.* It flows to cortical language circuits, making possible external linguistic communication about the contents of the narrative. The second is the *distributary of action.* It controls goal-directed behavior by way of connections with the basal ganglia and cortical motor circuits. And the third stream is the *distributary of consciousness,* the ground zero of explicit conscious experiences. Unlike the other two streams, the *distributary of consciousness* remains within the confines of PFC working memory, where it populates a second mental model, a conscious one, with explicit content in the form of a conscious higher-order state.

What I have described is a mental model account of Rosenthal's idea that we are conscious of the PFC higher-order state only if it is further re-represented / re-described by an additional higher-order state. That is what the second mental model does—it allows a conscious experience of the content of the pre-conscious mental model. To be conscious of the content of the second mental model, in Rosenthal's theory, would require yet another mental model, and on and on. But this infinite chain of additional re-representations / re-descriptions is not required. The conscious mental model feeds back to the pre-conscious model, which feeds forward to the conscious model. This loop allows the second mental model to be conscious of a version its own content, one that will have been updated by lower-order inputs to the pre-conscious model.

Matthias Michel suggested I call this the HO HO theory! In HO HO, the lower-order states are a neurobiological realm process, the first higher-order state is a cognitive realm process, and the second higher-order state is a conscious realm process.

Each distributary is a neural pathway that carries the mentalese narrative to its target circuit. Because each is a separate neural pathway that processes the narrative signal in its own way, the content reaching the targets can vary somewhat. For example, because verbal expression and overt action are different consequences of the pre-conscious mentalese narrative, what we say and what we do can be somewhat discordant. This may also explain why verbal reports, though fairly reliable as measures of momentary conscious experiences, do not always perfectly reflect what one experiences—we can't always put into words everything we are conscious of. It may also help us understand why, when stimuli are degraded or otherwise difficult to process (such as when they occur in peripheral vision), reports can be incomplete.

In this dual mental-model hypothesis, explicit consciousness of complex events emerges from interactions between granular and sub-granular PFC states. Lower-order non-PFC states, while often involved as inputs to the PFC, are not necessary for such higher-order conscious experiences. In other words, a thought, which is a higher-order state constructed by a pre-conscious mental model, is sufficient to populate the conscious higher-order state via the second mental model. As such, this is a HO HO HOROR version of HOT (sorry, I couldn't resist putting this in).

Figure 26.2 proposes that the output of the conscious mental model, much like the output of the pre-conscious mental model, is an abstract mentalese narrative (albeit a conscious one) that feeds distributaries flowing to motor circuits that control overt behavior and verbal expression. This implies that we have conscious agency, which you may know of as free will. The question of whether we actually make conscious choices is a matter of debate, with some arguing that

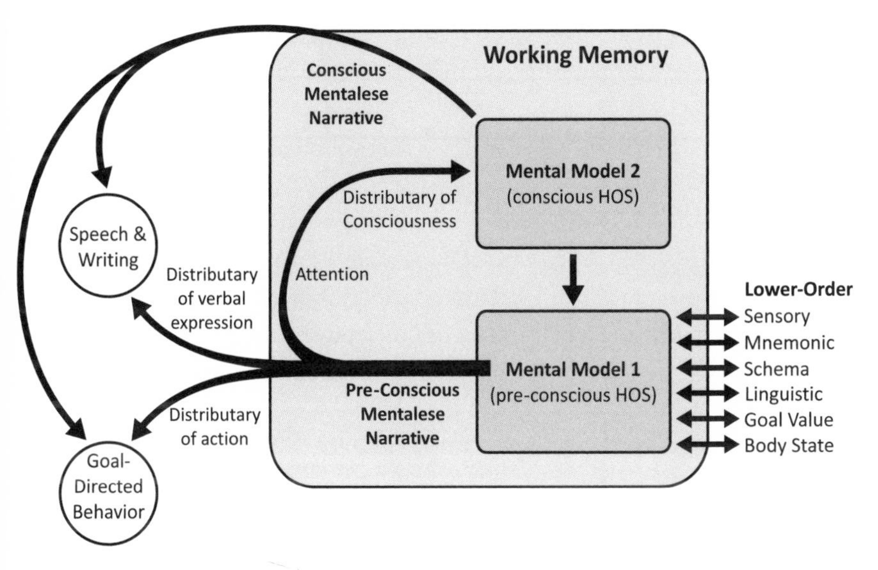

Figure 26.2. Conscious and pre-conscious mental models and their mentalese narratives control both verbal reporting and goal-directed behavior

our sense of choosing is illusory. But even with free will, we may sometimes wrongly attribute conscious control to our behaviors. This could happen when the conscious mental model notices responses that it did not control, and then generates a rationalization to explain the behavior (again the studies of split-brain patient Paul are relevant). In a classic article in 1978, right around the time of the studies of Paul, Richard Nisbett and Timothy Wilson referred to such rationalizations as instances of "telling more than we can know."

Cognition, Consciousness, and Energy

In Part IV, I described how cognitive realm events are more energy-demanding than neurobiological realm events. Similarly, constructing a conscious mental state requires more neural resources and energy than does the creation of non-conscious cognitive (including pre-conscious) states. Because less energy is required to sustain the pre-conscious mental model, it is the go-to mechanism for behavioral control in routine situations. The conscious model is engaged only when distinct advantages justify its energy expenditure.

But different kinds of conscious states also have different energy needs. The default condition for consciousness is low-energy tacit anoesis, which uses procedural meta-cognition to implicitly conceptualize lower-order states and maintain fringe experiences in the conscious mental model. Higher-energy noetic consciousness is called on when explicit factual or conceptual content is needed to explicitly categorize and conceptualize external or internal events, but also to verify anoetic feelings of rightness, or cope with feelings of wrongness. Super-high-energy autonoetic consciousnesses is engaged when, and only when, episodic memory is deployed to explicitly conceptualize your personal involvement in a situation, as in mental time travel.

Autonoesis, though energetically expensive, pays its way. It provides novel elements of mental modeling, enabling unparalleled degrees of flexible behavioral control and environmental mastery that go well beyond what can be done noetically or anoetically, or with non-conscious (including pre-conscious) cognitive control. Consistent with this idea are results from studies showing that areas involved in episodic memory, and hence possibly autonoesis, are implicated in the flexible control of behavior.

Knowing the Complications

The dual-mental-model idea depicted in Figures 26.1 and 26.2 is not to be confused with the popular two-system view of cognition discussed in Part IV. Recall that I re-organized the processes of the traditional system hypothesis, distributing them among three systems. My System 1 processes are neither cognitive nor conscious; my System 2 processes are cognitive, but not conscious; and my

System 3 processes are both cognitive and conscious. Consequently, in my scheme, model-free processes (System 1) are segregated from model-based processes (System 2 and System 3). But pre-conscious model-based processes (System 2) and conscious model-based processes (System 3) are also segregated from one another. The net result is that my three systems correspond to the neurobiological (System 1), cognitive (System 2), and conscious (System 3) realms of existence.

If I am correct that pre-conscious and conscious mental models can separately control overt behavior and verbal expression, the effort to scientifically understand consciousness may be even more complicated than we thought. Why? Because it would mean that we would not know, in a given moment, which mental model is in charge of what one says and/or does. This is indeed a complication, a challenge, for experiments on consciousness. But there may be an upside. Knowing what the complications are might help us to construct theories in which they are features rather than impediments. Paraphrasing Sun Tzu in *The Art of War,* you have to know your enemy to win the battle. Indeed, knowing that verbal reports and behavior can be controlled both consciously and pre-consciously may help explain why scientific solutions to the nature of consciousness seem so elusive. It might also help explain why it is so difficult to resolve debates about whether our sense that we consciously control our actions is illusory or real. In the perspective here, the sense of conscious agency that comes with responses produced by the pre-conscious mental model would be illusory, but actual conscious agency would be involved when the conscious mental model controls the responses.

The philosopher Jorge Morales recently questioned the extent to which behavior is controlled non-consciously. He suggested that because our introspections about our conscious experiences are so strong, we should resist thinking that complex behavioral control can occur non-consciously. But pre-conscious behavioral control is not the same as unconscious control. If my dual mental model turns out to be correct, he may have to re-think his position and allow both pre-conscious and conscious control of behavior.

After a recent lecture at Yale about some of the ideas in this book, a graduate student, Joan Ongchoco, sent me a suggestion about how pre-conscious and conscious verbal reports might be distinguished empirically:

> One possibility is that the difference between verbal reports controlled by the two mental models lies not just in the superficial words (i.e., what people say), but in the relations between these words (i.e., how people move from one idea to another). This is a process known as "semantic progression." The more implicit, "spontaneously" constructed, verbal reports that arise from a pre-conscious state may involve more "jumping" from idea to idea, compared to more "intentionally" constructed reports arising from the conscious model. Semantic progression can actually be measured through existing natural language processing models (e.g., latent semantic analysis or unsupervised embedders) that compute the semantic similarity between words or sentences (e.g., words in the same paragraph should show greater semantic similarity than words across two different paragraphs). Indeed, these types of analyses have been used to explore the progression of ideas in narrative texts and in film and might be leveraged to distinguish reports based on preconscious versus conscious states of mind.

If Ongchoco's fascinating and informative suggestion pans out, the metrics she suggested could be a way to search for neural networks in the brain that underlie the pre-conscious and conscious mental models.

Minds and Machines: The Sniff Test

If consciousness is just something that neural circuits do, it would seem that making computers "self-aware," autonoetic agents would be just a matter of figuring out the autonoetic circuitry in detail, duplicating its connectivity digitally, and querying the resulting machine about its conscious states. But I'm wary of this *strong AI* logic, which seems to take the so-called Turing test too seriously.

Proposed in 1950 by Alan Turing, the idea behind the Turing test was that if you cannot tell the difference between responses answered by a computer or a person, then the computer could be thought of as having intelligence. The strong AI view substitutes "consciousness" for "intelligence." But for me, this sleight of hand fails the sniff test. As the philosopher John Searle pointed out in 1980, machines can act intelligently without being conscious. Consciousness, he argued, requires the neurobiological mechanisms of a brain, a position he referred to as *biological naturalism*. More recently, the philosopher Tim Bayne proposed something similar, offering a *natural kind* of approach that treats consciousness as similar to digestion or respiration. So long as we don't interpret this as a defense of embodied consciousness, I think it's on the right track. The point is that consciousness is just as physical as digestion, respiration, reflex actions, conditioned responses, and habits, and also as physical as pre-conscious cognition. In other words, consciousness is, like these other processes, a legitimate topic for the natural sciences.

If we can't definitively tell from behavioral outputs whether an animal is responding consciously or non-consciously, how could we possibly use the output of a computer to tell if it is conscious? At least in animals, especially mammals, and most especially non-human primates, we have reason to suspect that their conscious circuitry was the foundation from which ours evolved (anoesis in early mammals paved the way for noesis in primates, and for autonoesis in humans, and possibly other great apes). While there is a good scientific reason for assuming that anoetic consciousness might be present in some form in all mammals, and noetic consciousness in all primates, actually demonstrating this is a different matter (see Chapter 25).

It took hundreds of millions of years for nervous systems to evolve the autonoetic kind of circuit we have. Knowing the circuit particulars is essential for understanding how the brain makes us conscious, but would not be sufficient to recreate reflective "self"-awareness (autonoetic consciousness) in a non-biological physical system. Underlying the evolution of our consciousness circuitry are eons of unknown microscopic cellular, molecular, and genetic adaptations. To recreate autonoesis, we might have to not only identify and recreate each of the microscopic changes, many of which were mere accidental events that had useful consequences, but also induce each change in a perfect recreation of the physical and social environments in which the changes occurred. Additionally, it might be necessary to factor in the impact of feedback to the nervous system from behavioral responses evolving in response to these dynamic environmental conditions at each step along the way.

Even if all this were possible, the question that would likely remain unanswerable is whether, to use the now trite expression, "there is any 'there' there." In other words, would a perfect computer simulation or robot actually have conscious states of the kind that developed through our biological, neurobiological, cognitive, and conscious evolution, or would it just be capable of generating a sophisticated set of responses that resemble those a human would make?

My take is that precise implementation in a non-biological physical system of all the neurobiological and cognitive processes that constitute what we think of as human consciousness would not be sufficient. I might be wrong, but as I see things today, human consciousness is only implementable in biological beings with our particular evolutionary history. The same holds for any and all other creatures that have, or that may possess, some form of consciousness.

We Are Our Neurons, and Our Neurons Are Ours

A major theme of this book has been that we are hierarchically organized biological units that, for the most part, function as an integrated system. Writing about such systems, the social scientist Donald Campbell noted in 1974 that processes at lower levels of a hierarchy are constrained by the workings of the higher level. Not controlled or determined by, but instead, constrained by. In the brain, conscious processes are higher than pre-conscious cognitive processes; these are higher than other non-cognitive processes that remain isolated from consciousness; and these, in turn, are higher than mere neurobiological processes.

Mike Gazzaniga offered this analogy to help understand how neural constraints work: just as cars are constrained by the traffic they create, the neural processes underlying mental processes are constrained by the mental processes they create. We can unpack this by saying that processes within each of our own unique hierarchical realms of existence depend on and constrain processes lower in the hierarchy. Ultimately, then, you are responsible—morally and legally—for what the totality of you, the organism, does. There are, of course, exceptions (including severe mental illness or loss of mental-state ownership). These, at least at this point, should be dealt with on a case-by-case basis, rather than by calling on stock defenses based on brain areas and / or non-conscious actions. In the final analysis, we are our neurons, but also, our neurons are ours—they reflect the individual and and collective histories of our four realms of existence.

Alfred Lotka, a scientific polymath in the early twentieth century, noted: "To say that a necessary condition for the writing of these words is the willing of the author to write them, and to say that a necessary condition for the writing of them is a certain state and configuration of the material of his brain . . . are probably merely two ways of saying the same thing." Rephrasing Lotka, I would say that consciousness and neural activity in the circuits that underlie it are two ways of describing what those circuits do. But without the psychological descriptions that we humans extract from our folk-psychological understanding of our introspections, it would be impossible to know what brain circuits do. It literally takes a mind to know what a mind is.

Closing the Circle

Back in the 1970s, when split-brain patient Paul—or rather, Paul's left hemisphere—confabulated without missing a beat an explanation for why a behavior it did not initiate was performed, Mike Gazzaniga and I concluded that these confabulations were not consciously generated concoctions. They seemed to flow freely from some non-conscious corner of his mind, and in so doing, rationalized away the momentary discordance brought on by the rogue behavior.

In light of the ideas discussed earlier, I can now suggest what might have been going on in Paul's brain. When we asked him why he did what he did, his momentary autonoetic awareness was colored with anoetic dissonance (a feeling of wrongness) since he (his talking left hemisphere) did not generate the response. In a flash, a pre-conscious mental model emerged, spun a mentalese narrative that flowed to his speech-control processes, and an explanation rolled off his tongue. Simultaneously, the mentalese narrative spawned a conscious mental model that interpreted his speech act as an adequate explanation of why he did what he did, thereby reducing the unsettling feeling of dissonance and making him consciously feel "right."

Why does the human brain spin such fables? Daniel Dennett offered an answer, suggesting that our narratives are a defense tactic, a way of defining and protecting our understanding of our self. Similarly, Julia Hailova and colleagues proposed that we adjust appraisals (narratives) of our past and future self to maintain a favorable view of our present self. When Dennett, Gallagher, Flanagan, Bruner, Gazzaniga, I, and others linked self, consciousness, and narratives, we were talking about autonoetic consciousness, whether or not we used the term autonoesis (and usually we did not).

Sigmund Freud is the subject of ridicule in contemporary philosophy, psychology, and neuroscience for his theories, including his theory of unconscious ego-defense mechanisms. But the idea that narratives flowing from a "non-conscious" mental model help us to "defend" our autonoetic understanding of who we are (our ego) likely has Freud chuckling in his grave.

That the neural basis of those pre-conscious cognitive and memory processes underlying anoetic, noetic, and autonoetic consciousness are known, or at least knowable, takes us all the way to the last microsecond of pre-conscious processing—in other words, all the way to the consciousness finishing line. The mentalese narrative does the rest.

This hypothetical conception will likely raise philosophical and scientific eyebrows. But regardless of whether the mentalese narrative part of the hypothesis turns out to be correct, the idea that our conscious experiences are complex amalgams constructed by pre-conscious and conscious mental models is one grounded in the neuroscience of behavior and cognition, and it provides a novel way to think about, and research, consciousness.

Although we still have much to learn about consciousness, a good conception is always the best road to scientific understanding. While I think that the conception I have developed here is pretty good, time will be the judge. In the meantime, I'm going to continue down this road and see what happens along the way. Join at your own risk!

Selected Sources and Further Reading

The literature on which this book draws is extensive. Full citations of the works referred to in this book can be found at joseph-ledoux.com/The-Four-Realms-of-Existence. For those who wish to read more about the topics I discuss here, please see below.

Introduction

Gazzaniga, M. S. *The Bisected Brain*. New York: Appleton-Century-Crofts, 1970.

———. "One Brain—Two Minds?" *American Scientist* 60 (1972): 311–317.

Gazzaniga, M. S., and J. E. LeDoux. *The Integrated Mind*. New York: Plenum, 1978.

LeDoux, J. *The Deep History of Ourselves: The Four-Billion-Year Story of How We Got Conscious Brains*. New York: Viking, 2019.

LeDoux, J. E., D. H. Wilson, and M. S. Gazzaniga. "A Divided Mind: Observations on the Conscious Properties of the Separated Hemispheres." *Annals of Neurology* 2 (1978): 417–421.

Chapter 1

Boring, E. G. *A History of Experimental Psychology*. New York: Appleton-Century-Crofts, 1950.

Hall, C. S., G. Lindzey, and J. B. Campbell. *Theories of Personality.* New York: John Wiley & Sons, 1998.

Hyman, S. E. "Revolution Stalled." *Science of Translational Medicine* 4 (2012): 155cm111.

Insel, T., B. Cuthbert, M. Garvey, R. Heinssen, D. S. Pine, K. Quinn, C. Sanislow, and P. Wang. "Research Domain Criteria (RDoC): Toward a New Classification Framework for Research on Mental Disorders." *American Journal of Psychiatry* 167 (2010): 748–751.

Schneider, S. *Artificial You: AI and the Future of Your Mind.* Princeton, NJ: Princeton University Press, 2019.

Taschereau-Dumouchel, V., M. Michel, H. Lau, S. G. Hofmann, and J. E. LeDoux. "Putting the 'Mental' Back in 'Mental Disorders': A Perspective from Research on Fear and Anxiety." *Molecular Psychiatry* 27 (2022): 1322–1330.

Chapter 2

Chalmers, D. "Strong and Weak Emergence," in P. Clayton and P. Davies, eds., *The Re-Emergence of Emergence: The Emergentist Hypothesis from Science to Religion.* Oxford Academic, May 2008. https://doi.org/10.1093/acprof:oso/9780199544318.003.0011, accessed October 9, 2022.

Danziger, K. "The Historical Formation of Selves." Pp. 137–159 in R. D. Ashmore and L. J. Jussim, eds., *Self and Identity: Fundamental Issues.* New York: Oxford University Press, 1997.

Dennett, D.C. "The Self as a Center of Narrative Gravity." Pp. 103–115 in F. Kessel et al., eds., *Self and Consciousness: Multiple Perspectives.* Hillsdale, NJ: Erlbaum, 1992.

Flanagan, O. "Neuroscience: Knowing and Feeling." *Nature* 469 (2011): 160–161.

Gallagher, S. "Philosophical Conceptions of the Self: Implications for Cognitive Science." *Trends in Cognitive Sciences* 4 (2000): 14–21.

Gardner, H. *The Mind's New Science: A History of the Cognitive Revolution.* New York: Basic Books, 1987.

Hall, C. S., G. Lindzey, and J. B. Campbell. *Theories of Personality.* New York: John Wiley & Sons, 1998.

James, W. *Principles of Psychology.* New York: Holt, 1890.

Metzinger, T. *Being No One.* Cambridge, MA: MIT Press, 2003.

Chapter 3

Damasio, A. *Self Comes to Mind: Constructing the Conscious Brain.* New York: Pantheon, 2010.

Hall, C. S., G. Lindzey, and J. B. Campbell. *Theories of Personality.* New York: John Wiley & Sons, 1998.

Johnson, M. "What Makes a Body?" *Journal of Speculative Philosophy* 22 (2008): 159–169.

LeDoux, J. E. *Synaptic Self: How Our Brains Become Who We Are.* New York: Viking, 2002.

Metzinger, T. "First-Order Embodiment, Second-Order Embodiment, Third-Order Embodiment: From Spatiotemporal Self-Location to Minimal Selfhood." Pp. 272–286 in R. Shapiro, ed., *The Routledge Handbook of Embodied Cognition.* New York: Routledge, 2014.

Northoff, G. P., T. Qin, and E. Feinberg. "Brain Imaging of the Self-Conceptual, Anatomical and Methodological Issues." *Consciousness and Cognition* 20 (2011): 52–63.

Panksepp, J. *Affective Neuroscience.* New York: Oxford University Press, 1998.

Chapter 4

Brick, C., B. Hood, V. Ekroll, and L. de-Wit. "Illusory Essences: A Bias Holding Back Theorizing in Psychological Science." *Perspectives on Psychological Science* 17, no. 2 (2021): 491–506.

Bruner, J. "The 'Remembered' Self." Pp. 41–54 in R. Fivush and U. Neisser, eds., *The Remembering Self: Construction and Accuracy in the Self-Narrative.* Cambridge, UK: Cambridge University Press, 1994.

Danziger, K. *Naming the Mind: How Psychology Found Its Language.* London: Sage, 1997.

LeDoux, J. E. "Semantics, Surplus Meaning, and the Science of Fear." *Trends in Cognitive Sciences* 21 (2017): 303–306.

Mandler, G., and W. Kessen. *The Language of Psychology.* New York: John Wiley & Sons, 1959.

Schaffner, K. F. *Construct Validity in Psychology and Psychiatry.* Submitted for publication.

Chapter 5

Bruner, J. "The 'Remembered' Self." Pp. 44–54 in R. Fivush and U. Neisser, eds., *The Remembering Self: Construction and Accuracy in the Self-Narrative.* Cambridge, UK: Cambridge University Press, 1994.

Dennett, D. C. *Darwin's Dangerous Idea.* New York: Simon and Schuster, 1995.

———. "The Self as a Center of Narrative Gravity." Pp. 103–115 in F. Kessel et al., eds., *Self and Consciousness: Multiple Perspectives.* Hillsdale, NJ: Erlbaum, 1992.

Fireman, G. D., T. E. McVay, and O. J. Flanagan, eds. *Narrative and Consciousness: Literature, Psychology, and the Brain.* Oxford, UK: Oxford University Press, 2003.

Gazzaniga, M. S. *The Social Brain.* New York: Basic Books, 1985.

Ginsburg, S., and E. Jablonka. *The Evolution of the Sensitive Soul.* Cambridge, MA: MIT Press, 2019.

Metzinger, T. "First-Order Embodiment, Second-Order Embodiment, Third-Order Embodiment: From Spatiotemporal Self-Location to Minimal Selfhood." Pp. 272–286 in R. Shapiro, ed., *The Routledge Handbook of Embodied Cognition.* New York: Routledge, 2014.

Volk, T. *Quarks to Culture: How We Came to Be.* New York: Columbia University Press, 2017.

Chapter 6

Alanen, L. *Descartes's Concept of Mind.* Cambridge, MA: Harvard University Press, 2003.

Bechtel, W., and R. C. Richardson. "Vitalism." In E. Craig., ed., *Routledge Encyclopedia of Philosophy,* vol. 9. London: Routledge, 1998.

Bernard, C. *An Introduction to the Study of Experimental Medicine.* New York: Collier, 1961.

Cannon, W. B. *Bodily Changes in Pain, Hunger, Fear, and Rage.* New York: Appleton, 1929.

Garrett, B. J. "What the History of Vitalism Teaches Us about Consciousness and the 'Hard Problem.'" *Philosophy and Phenomenological Research* 72 (2006): 576–588.

Haigh, E. "The Roots of the Vitalism of Xavier Bichat." *Bulletin of the History of Medicine* 49 (1975): 72–86.

Shields, C. "Aristotle's Psychology." In E. N. Zalta, ed., *The Stanford Encyclopedia of Philosophy,* Metaphysics Research Lab, Stanford University, 2020. https://plato.stanford.edu/archives/win2020/entries/aristotle-psychology, accessed February 24, 2021.

Chapter 7

Buss, L. W. *The Evolution of Individuality.* Princeton, NJ: Princeton University Press, 1987.

Dawkins, R. *The Selfish Gene.* New York: Oxford University Press, 1976.

Godfrey-Smith, P. *Darwinian Populations and Natural Selection.* Oxford, UK: Oxford University Press, 2009.

Hull, D. "Individual." In E. F. Keller and E. A. Lloyd, eds., *Keywords in Evolutionary Biology.* Cambridge, MA: Harvard University Press, 1992.

Lewontin, R. C. "The Units of Selection." *Annual Review of Ecology and Systematics* 1 (1970): 1–18.

Maturana, H. R., and F. J. Varela. *The Tree of Knowledge: The Biological Roots of Human Understanding.* Boston: New Science Library, 1987.

Pradeu, T. "What Is an Organism? An Immunological Answer." *History and Philosophy of the Life Sciences* 32 (2010): 247–267.

Wilson, R. A., and M. J. Barker. "Biological Individuals." In E. N. Zalta, ed., *The Stanford Encyclopedia of Philosophy,* Metaphysics Research Lab, Stanford University, 2019.

Chapter 8

Romer, A. S. "The Vertebrate as a Dual Animal—Somatic and Visceral." Pp. 121–156 in T. Dobzhansky et al., eds., *Evolutionary Biology,* vol. 6 (New York: Springer, 1972).

———. "The Vertebrate as a Dual Animal—Visceral and Somatic." *Anatomical Record* 132 (1958): 496.

———. *Vertebrate Body.* Philadelphia: W.B. Saunders, 1955.

Chapter 9

Ginsburg, S., and E. Jablonka. "The Evolution of Associative Learning: A Factor in the Cambrian Explosion." *Journal of Theoretical Biology* 266 (2010): 11–20.

Hills, T. T. "Animal Foraging and the Evolution of Goal-Directed Cognition." *Cognitive Science* 30 (2006): 3–41.

LeDoux, J. *The Deep History of Ourselves: The Four-Billion-Year Story of How We Got Conscious Brains.* New York: Viking, 2019.

Chapter 10

Cisek, P. "Evolution of Behavioural Control from Chordates to Primates." *Philosophical Transactions of the Royal Society London B Biological Sciences* 377, no. 1844 (2022): 20200522.

Kaas, J. H., H. X. Qi, and I. Stepniewska. "Escaping the Nocturnal Bottleneck, and the Evolution of the Dorsal and Ventral Streams of Visual Processing in Primates." *Philosophical Transactions of the Royal Society London B Biological Sciences* 377, no. 1844 (2022): 20210293.

LeDoux, J. *The Deep History of Ourselves: The Four-Billion-Year Story of How We Got Conscious Brains.* New York: Viking, 2019.

MacLean, P. D. "Some Psychiatric Implications of Physiological Studies on Frontotemporal Portion of Limbic System (Visceral Brain)." *Electroencephalography and Clinical Neurophysiology* 4 (1952): 407–418.

———. *The Triune Brain in Evolution: Role in Paleocerebral Functions.* New York: Plenum, 1990.

Striedter, G. F., and R. G. Northcutt. *Brains through Time: A Natural History of Vertebrates.* New York: Oxford University Press, 2020.

Chapter 11

Arendt, D., M. A. Tosches, and H. Marlow. "From Nerve Net to Nerve Ring, Nerve Cord and Brain—Evolution of the Nervous System." *Nature Reviews Neuroscience* 17 (2016): 61–72.

Holland, N. D. "Early Central Nervous System Evolution: An Era of Skin Brains?" *Nature Reviews Neuroscience* 4 (2003): 617–627.

Romer, A. S. "The Vertebrate as a Dual Animal—Somatic and Visceral." Pp. 121–156 in T. Dobzhansky et al., eds., *Evolutionary Biology,* vol. 6 (New York: Springer, 1972).

———. "The Vertebrate as a Dual Animal—Visceral and Somatic." *Anatomical Record* 132 (1958): 496.

Chapter 12

Bichat, X. *Physiological Researches on Life and Death.* New York: Arno Press, 1977.

Blessing, B., and I. Gibbins. "Autonomic Nervous System." *Scholarpedia* 3 (2008): 2787.

Blessing, W. W. "Inadequate Frameworks for Understanding Bodily Homeostasis." *Trends in Neurosciences* 20 (1997): 235–239.

Cannon, W. B. *Bodily Changes in Pain, Hunger, Fear, and Rage.* New York: Appleton, 1929.

Gibbins, I. "Functional Organization of Autonomic Neural Pathways." *Organogenesis* 9 (2013): 169–175.

Chapter 13

Dezfouli, A., and B. W. Balleine. "Habits, Action Sequences, and Reinforcement Learning." *European Journal of Neuroscience* 35 (2012): 1036–1051.

Fanselow, M. S., and L. S. Lester. "A Functional Behavioristic Approach to Aversively Motivated Behavior: Predatory Imminence as a Determinant of the Topography of Defensive Behavior." Pp. 185–211 in R. C. Bolles and M. D. Beecher, eds., *Evolution and Learning.* Hillsdale, N.J.: Erlbaum, 1988.

Fanselow, M. S., and A. M. Poulos. "The Neuroscience of Mammalian Associative Learning." *Annual Review of Psychology* 56 (2005): 207–234.

Graybiel, A. M. "Habits, Rituals, and the Evaluative Brain." *Annual Review of Neuroscience* 31 (2008): 359–387.

LeDoux, J. E. *The Deep History of Ourselves: The Four-Billion-Year Story of How We Got Conscious Brains.* New York: Viking, 2019.

———. *The Emotional Brain.* New York: Simon and Schuster, 1996.

———. "Rethinking the Emotional Brain." *Neuron* 73 (2012): 653–676.

———. *Synaptic Self: How Our Brains Become Who We Are.* New York: Viking, 2002.

Poldrack, R. A. *Hard to Break: Why Our Brains Make Habits Stick.* Princeton, NJ: Princeton University Press, 2021.

Robbins, T. W., and R. M. Costa. "Habits." *Current Biology* 27 (2017): R1200–R1206.

Schroer, S. A. "Jakob von Uexküll: The Concept of Umwelt and Its Potentials for an Anthropology beyond the Human." *Ethnos* 86 (2021): 132–152.

Sherrington, C. S. *The Integrative Action of the Nervous System*. New Haven: Yale University Press, 1906.

Shettleworth, S. J. "Animal Cognition and Animal Behaviour." *Animal Behaviour* 61 (2001): 277–286.

Thorndike, E. L. "Animal Intelligence: An Experimental Study of the Associative Processes in Animals." *Psychological Monographs* 2 (1898): 109.

Chapter 14

Craik, K. J. W. *The Nature of Explanation*. Cambridge, UK: Cambridge University Press, 1943.

Hebb, D. O. *The Organization of Behavior.* New York: John Wiley and Sons, 1949.

Keller, F. S. *The Definition of Psychology.* New York: Appleton-Century-Crofts, 1973.

Lashley, K. "The Problem of Serial Order in Behavior." Pp. 112–146 in L. A. Jeffers, ed., *Cerebral Mechanisms in Behavior.* New York: Wiley, 1950.

Mandler, G. "Origins of the Cognitive (R)evolution." *Journal of the History of the Behavioral Sciences* 38 (2002): 339–353.

Miller, G. "The Magical Number Seven, Plus or Minus Two: Some Limits on Our Capacity for Processing Information." *Psychological Review* 63 (1956): 81–97.

Miller, G. A., E. G. Glanter, and K. H. Pribram. *Plans and the Structure of Behavior.* New York: Holt, Rinehart, and Winston, 1960.

Neisser, U. *Cognitive Psychology.* Englewood Cliffs, NJ: Prentice Hall, 1967.

Scoville, W. B., and B. Milner. "Loss of Recent Memory after Bilateral Hippocampal Lesions." *Journal of Neurology and Psychiatry* 20 (1957): 11–21.

Tolman, E. C. "Cognitive Maps in Rats and Men." *Psychological Review* 55 (1948): 189–208.

Chapter 15

Baddeley, A. "The Episodic Buffer: A New Component of Working Memory?" *Trends in Cognitive Sciences* 4 (2000): 417–423.

Baddeley, A., and G. J. Hitch. "Working Memory." Pp. 47–89 in T. Bower, ed., *The Psychology of Learning and Motivation,* vol. 8. New York: Academic Press, 1974.

Dickinson, A., and B. W. Balleine. "Motivational Control of Goal-Directed Action." *Animal Learning and Behavior* 22 (1994).

Evans, J. S., and K. E. Stanovich. "Dual-Process Theories of Higher Cognition: Advancing the Debate." *Perspectives on Psychological Science* 8 (2013): 223–241.

Fuster, J. *The Prefrontal Cortex*. New York: Academic Press, 2008.

Goldman-Rakic, P. S. "Circuitry of Primate Prefrontal Cortex and Regulation of Behavior by Representational Memory." Pp. 373–418 in F. Blum, ed., *Handbook of Physiology,* vol. 5: *Higher Functions of the Brain*. Bethesda, MD: American Physiological Society, 1987.

Kahneman, D. *Thinking, Fast and Slow.* New York: Farrar, Straus, and Giroux, 2011.

Kihlstrom, J. F. "The Cognitive Unconscious." *Science* 237 (1987): 1445–1452.

Lundqvist, M., J. Rose, M. R. Warden, T. Buschman, E. K. Miller, and P. Herman. "A Hot-Coal Theory of Working Memory." bioRxiv (2021). https://doi.org/10.1101/2020.12.30.424833.

Miller, E. K., and J. D. Cohen. "An Integrative Theory of Prefrontal Cortex Function. *Annual Review of Neuroscience* 24 (2001): 167–202.

O'Keefe, J., and L. Nadel. *The Hippocampus as a Cognitive Map.* Oxford, UK: Clarendon Press, 1978.

Squire, L. *Memory and Brain*. New York: Oxford University Press, 1987.

Chapter 16

Daw, N. D., Y. Niv, and P. Dayan. "Uncertainty-Based Competition between Prefrontal and Dorsolateral Striatal Systems for Behavioral Control." *Nature Neuroscience* 8 (2005): 1704–1711.

Dijksterhuis, A., and L. F. Nordgren. "A Theory of Unconscious Thought." *Perspectives on Psychological Science* 1 (2006): 95–109.

Fleming, S. M. *Know Thyself: The Science of Self-Awareness.* New York: Basic Books, 2021.

Johnson-Laird, P. N. *Mental Models: Towards a Cognitive Science of Language, Inference, and Consciousness.* Cambridge, MA: Harvard University Press, 1983.

Kelley, C. M., and L. L. Jacoby. "Adult Egocentrism: Subjective Experience versus Analytic Bases for Judgment." *Journal of Memory and Language* 35 (1996): 157–175.

Koriat, A. "Metacognition and Consciousness." In E. Thompson et al., eds., *The Cambridge Handbook of Consciousness.* Cambridge, UK: Cambridge University Press, 2007.

LeDoux, J. *The Deep History of Ourselves: The Four-Billion-Year Story of How We Got Conscious Brains.* New York: Viking, 2019.

LeDoux, J., and N. D. Daw. "Surviving Threats: Neural Circuit and Computational Implications of a New Taxonomy of Defensive Behaviour." *Nature Reviews Neuroscience* 19 (2018): 269–282.

Melloni, L. "Consciousness as Interference in Time: A Commentary on Victor Lamme." Pp. 881–893 in T. Metzinger and J. M. Windt, eds., *Open MIND.* Cambridge, MA: MIT Press, 2015.

Reber, T. P., R. Luechinger, P. Boesiger, and K. Henke. "Unconscious Relational Inference Recruits the Hippocampus." *Journal of Neuroscience* 32 (2012): 6138–6148. PMC6622124.

Seth, A. *Being You: A New Science of Consciousness.* New York: Penguin Random House, 2021.

Strick, M., A. Dijksterhuis, M. W. Bos, A. Sjoerdsma, R. B. van Baaren, and L. F. Nordgren. "A Meta-Analysis on Unconscious Thought Effects." *Social Cognition* 29 (2011): 738–762.

Chapter 17

Abramson, C. I., and H. Wells. "An Inconvenient Truth: Some Neglected Issues in Invertebrate Learning." *Perspectives on Psychological Science* 41 (2018): 395–416. PMC6701716.

Bennett, M. S. "Five Breakthroughs: A First Approximation of Brain Evolution from Early Bilaterians to Humans." *Frontiers in Neuroanatomy* 15 (2021): 693346. PMC8418099.

Chittka, L. *The Mind of a Bee.* Princeton, NJ: Princeton University Press, 2022.

Clayton, N. S., and A. Dickinson. "Episodic-like Memory during Cache Recovery by Scrub Jays." *Nature* 395 (1998): 272–274.

Clayton, N. S., D. P. Griffiths, N. J. Emery, and A. Dickinson. "Elements of Episodic-like Memory in Animals." *Philosophical Transactions of the Royal Society London B Biological Sciences* 356 (2001): 1483–1491. PMC1088530.

Giurfa, M. "Social Learning in Insects: A Higher-Order Capacity?" *Frontiers in Behavioral Neuroscience* 6 (2012): 57. PMC3433704.

Godfrey-Smith, P. *Metazoa: Animal Life and the Birth of the Mind.* New York: Farrar, Straus, and Giroux, 2020.

LeDoux, J. *The Deep History of Ourselves: The Four-Billion-Year Story of How We Got Conscious Brains.* New York: Viking, 2019.

LeDoux, J. E. "What Emotions Might Be Like in Other Animals." *Current Biology* 31 (2021): R824–R829.

Murray, E. A., S. P. Wise, and K. S. Graham. *The Evolution of Memory Systems: Ancestors, Anatomy, and Adaptations.* Oxford, UK: Oxford University Press, 2017.

Perry, C. J., A. B. Barron, and K. Cheng. "Invertebrate Learning and Cognition: Relating Phenomena to Neural Substrate." *Wiley Interdisciplinary Reviews: Cognitive Science* 4 (2013): 561–582.

Stone, T., B. Webb, A. Adden, N. B. Weddig, A. Honkanen, R. Templin, W. Wcislo, L. Scimeca, E. Warrant, and S. Heinze. "An Anatomically Constrained Model for Path Integration in the Bee Brain." *Current Biology* 27 (2017): 3069–3085, e3011. PMC6196076.

Chapter 18

Corballis, M. C. "The Uniqueness of Human Recursive Thinking." *American Scientist* 95 (2007): 240.

Dennett, D. C. *Kinds of Minds: Toward an Understanding of Consciousness.* New York: Basic Books, 1996.

Hills, T. T., P. M. Todd, and R. L. Goldstone. "Search in External and Internal Spaces: Evidence for Generalized Cognitive Search Processes." *Psychological Science* 19 (2008): 802–808.

Kaas, J. H., H. X. Qi, and I. Stepniewska. "Escaping the Nocturnal Bottleneck, and the Evolution of the Dorsal and Ventral Streams of Visual Processing in Primates." *Philosophical Transactions of the Royal Society London B Biological Sciences* 377 (2022): 20210293. PMC8710890.

Murray, E. A., S. P. Wise, and K. S. Graham. *The Evolution of Memory Systems: Ancestors, Anatomy, and Adaptations.* Oxford, UK: Oxford University Press, 2017.

Penn, D. C., K. J. Holyoak, and D. J. Povinelli. "Darwin's Mistake: Explaining the Discontinuity between Human and Nonhuman Minds." *Behavioral and Brain Sciences* 31 (2008): 109–130; discussion 130–178.

Preuss, T. M. "Evolutionary Specializations of Primate Brain Systems." In M. J. Ravosa and M. Dagosto, eds., *Primate Origins: Adaptations and Evolution.* Boston: Springer, 2007.

Read, D. W. "Working Memory: A Cognitive Limit to Non-Human Primate Recursive Thinking Prior to Hominid Evolution." *Evolutionary Psychology* 6 (2008): 676–714.

Rudebeck, P. H., and A. Izquierdo. "Foraging with the Frontal Cortex: A Cross-Species Evaluation of Reward-Guided Behavior." *Neuropsychopharmacology* 47 (2022): 134–146. PMC8617092.

Schacter, D. L., and D. R. Addis. "On the Constructive Episodic Simulation of Past and Future Events." *Behavioral and Brain Sciences* 30 (2007): 331–332.

Seed, A., and M. Tomasello. "Primate Cognition." *Topics in Cognitive Science* 2 (2010): 407–419.

Suddendorf, T., J. Redshaw, and A. Bulley. *The Invention of Tomorrow.* New York: Basic Books, 2022.

Zamani, A., R. Carhart-Harris, and K. Christoff. "Prefrontal Contributions to the Stability and Variability of Thought and Conscious Experience." *Neuropsychopharmacology* 47 (2022): 329–348. PMC8616944.

Chapter 19

Balleine, B. W. "The Meaning of Behavior: Discriminating Reflex and Volition in the Brain." *Neuron* 104 (2019): 47–62.

Bramson, B., D. Folloni, L. Verhagen, B. Hartogsveld, R. B. Mars, I. Toni, and K. Roelofs. "Human Lateral Frontal Pole Contributes to Control over Emotional Approach-Avoidance Actions." *Journal of Neuroscience* 40 (2020): 2925–2934. PMC7117901.

Burgess, P. W., I. Dumontheil, and S. J. Gilbert. "The Gateway Hypothesis of Rostral Prefrontal Cortex (Area 10) Function." *Trends in Cognitive Sciences* 11 (2007): 290–298.

Gilboa, A., and H. Marlatte. "Neurobiology of Schemas and Schema-Mediated Memory." *Trends in Cognitive Sciences* 21 (2017): 618–631.

Jacobsen, C. F. "Studies of Cerebral Function in Primates. I. The Functions of the Frontal Associations Areas in Monkeys." *Comparative Psychology Monographs* 13 (1936): 3–60.

Kaas, J. H. "The Evolution of Brains from Early Mammals to Humans." *Wiley Interdisciplinary Reviews: Cognitive Science* 4 (2013): 33–45. PMC3606080.

Mannella, F., M. Mirolli, and G. Baldassarre. "Goal-Directed Behavior and Instrumental Devaluation: A Neural System-Level Computational Model." *Frontiers in Behavioral Neuroscience* 10 (2016): 181. PMC5067467.

Mansouri, F. A., E. Koechlin, M. G. P. Rosa, and M. J. Buckley. "Managing Competing Goals—A Key Role for the Frontopolar Cortex." *Nature Reviews Neuroscience* 18 (2017): 645–657.

Miller, E. K., and J. D. Cohen. "An Integrative Theory of Prefrontal Cortex Function." *Annual Review of Neuroscience* 24 (2001): 167–202.

Miller, E. K., M. Lundqvist, and A. M. Bastos. "Working Memory 2.0." *Neuron* 100 (2018): 463–475.

Passingham, R. E., J. B. Smaers, and C. C. Sherwood. "Evolutionary Specializations of the Human Prefrontal Cortex." Pp. 207–226 in J. H. Kaas and T. M. Preuss, eds., *Evolution of Nervous Systems.* New York: Elsevier, 2017.

Postle, B. R. "Working Memory as an Emergent Property of the Mind and Brain." *Neuroscience* 139 (2006): 23–38. PMC1428794.

Preuss, T. M. "The Human Brain: Rewired and Running Hot." *Annals of the New York Academy of Sciences* 1225, supp. 1 (2011): E182–191. PMC3103088.

Preuss, T. M., and S. P. Wise. "Evolution of Prefrontal Cortex." *Neuropsychopharmacology* 47 (2022): 3–19.

Rudebeck, P. H., and A. Izquierdo. "Foraging with the Frontal Cortex: A Cross-Species Evaluation of Reward-Guided Behavior. *Neuropsychopharmacology* 47 (2022): 134–146. PMC8617092.

Schacter, D. L., and D. R. Addis. "On the Constructive Episodic Simulation of Past and Future Events." *Behavioral and Brain Sciences* 30 (2007): 331–332.

Chapter 20

Bruner, J. "Life as Narrative." *Social Research* 54 (1987): 11–32.

Carnap, R. *The Unity of Science.* London: Kegan Paul, Trench, Trubner, and Co., 1934.

Chalmers, D. *The Conscious Mind.* New York: Oxford University Press, 1996.

Churchland, P. M. *Matter and Consciousness.* Cambridge, MA: MIT Press, 1984.

Lau, H., and M. Michel. "A Socio-Historical Take on the Meta-Problem of Consciousness." *Journal of Consciousness Studies* 26 (2019): 136–147.

Michel, M., et al. "Opportunities and Challenges for a Maturing Science of Consciousness." *Nature Human Behaviour* 3 (2019): 104–107. PMC6568255.

Nagel, T. "What Is It Like to Be a Bat?" *Philosophical Review* 83 (1974): 4435–4450.

Seth, A. *Being You: A New Science of Consciousness.* New York: Penguin Random House, 2021.

Tononi, G., and C. Koch. "Consciousness: Here, There and Everywhere?" *Philosophical Transactions of the Royal Society London B Biological Sciences* 370 (2015). PMC4387509.

Chapter 21

Baars, B. J. *A Cognitive Theory of Consciousness.* New York: Cambridge University Press, 1988.

Block, N. "Concepts of Consciousness." Pp. 206–218 in D. Chalmers, ed., *Philosophy of Mind: Classical and Contemporary Readings.* New York: Oxford University Press, 2002.

Brown, R. "The HOROR Theory of Phenomenal Consciousness." *Philosophical Studies* 172 (2015): 1783–1794.

Brown, R., Lau, H., and J. E. LeDoux. "Understanding the Higher-Order Approach to Consciousness." *Trends in Cognitive Sciences* 23 (2019): 754–768.

Changeux, J.-P., and L. Naccache. "The Global Neuronal Workspace Model of Conscious Access: From Neuronal Architectures to Clinical Applications." Pp. 55–84 in S. Dehaene and Y. Christen, eds., *Characterizing Consciousness: From Cognition to the Clinic?* Berlin: Springer Berlin Heidelberg, 2011.

Cleeremans, A., D. Achoui, A. Beauny, L. Keuninckx, J. R. Martin, S. Muñoz-Moldes, L. Vuillaume, and A. de Heering. "Learning to Be Conscious." *Trends in Cognitive Sciences* 24 (2020): 112–123.

Crick, F., and C. Koch. "The Problem of Consciousness." *Scientific American* 267 (1992): 152–159.

Dehaene, S., H. Lau, and S. Kouider. "What Is Consciousness, and Could Machines Have It?" *Science* 358 (2017): 486–492.

Fleming, S. M. "Awareness as Inference in a Higher-Order State Space." *Neuroscience of Consciousness* 1 (2020). PMC7065713.

Graziano, M. S. A. "Consciousness and the Attention Schema: Why It Has to Be Right." *Cognitive Neuropsychology* 37 (2020): 224–233.

Lau, H. "Consciousness, Metacognition, and Perceptual Reality Monitoring." *PsyArXiv Preprints,* 2019. https://doi.org/10.31234/osf.io/ckbyf.

———. *In Consciousness We Trust: The Cognitive Neuroscience of Subjective Experience.* New York. Oxford University Press, 2022.

Lau, H., and R. Brown. "The Emperor's New Phenomenology? The Empirical Case for Conscious Experience without First-Order Representations." Pp. 171–197 in A. Pautz and D. Stoljar, eds., *Blockheads! Essays on Ned Block's Philosophy of Mind and Consciousness.* Cambridge, MA: MIT Press, 2019.

LeDoux, J. E., M. Michel, and H. Lau. "A Little History Goes a Long Way toward Understanding Why We Study Consciousness the Way We Do Today." *Proceedings of the National Academy of Sciences USA* 117 (2020): 6976–6984. PMC7132279.

Morales, J., H. Lau, and S. M. Fleming. "Domain-General and Domain-Specific Patterns of Activity Supporting Metacognition in Human Prefrontal Cortex." *Journal of Neuroscience* 38 (2018): 3534–3546. PMC5895040.

Rosenthal, D. M. *Consciousness and Mind.* Oxford, UK: Oxford University Press, 2005.

Rosenthal, D., and J. Weisberg. "Higher-Order Theories of Consciousness." *Scholarpedia* 3 (2008): 4407.

Sergent, C., M. Corazzol, G. Labouret, F. Stockart, M. Wexler, J. R. King, F. Meyniel, and D. Pressnitzer. "Bifurcation in Brain Dynamics Reveals a Signature of Conscious Processing Independent of Report." *Nature Communications* 12 (2021): 1149. PMC7895979.

Shea, N., and C. D. Frith. "The Global Workspace Needs Metacognition." *Trends in Cognitive Sciences* 23 (2019): 560–571.

Chapter 22

Brown, R., H. Lau, and J. E. LeDoux. "Understanding the Higher-Order Approach to Consciousness." *Trends in Cognitive Sciences* 23 (2019): 754–768.

Dijkstra, N., P. Kok, and S. M. Fleming. "Perceptual Reality Monitoring: Neural Mechanisms Dissociating Imagination from Reality." *Neuroscience and Biobehavioral Reviews* 135 (2022): 104557.

Friston, K. J., and C. D. Frith. "Active Inference, Communication and Hermeneutics." *Cortex* 68 (2015): 129–143. PMC4502445.

Lau, H., M. Michel, J. E. LeDoux, and S. M. Fleming. "The Mnemonic Basis of Subjective Experience." *Nature Reviews Psychology* 1 (2022): 479–488. https://doi.org/10.1038/s44159-022-00068-6.

LeDoux, J. *The Deep History of Ourselves: The Four-Billion-Year Story of How We Got Conscious Brains.* New York: Viking, 2019.

LeDoux, J. E., and H. Lau. "Seeing Consciousness through the Lens of Memory." *Current Biology* 30 (2020): R1018–R1022.

Michel, M., and J. Morales. "Minority Reports: Consciousness and the Prefrontal Cortex." *Mind and Language* 35 (2020): 493–513.

Odegaard, B., R. T. Knight, and H. Lau. "Should a Few Null Findings Falsify Prefrontal Theories of Conscious Perception?" *Journal of Neuroscience* 37 (2017): 9593–9602. PMC5628405.

Schacter, D., and D. Addis. "Memory and Imagination: Perspectives on Constructive Episodic Simulation." Pp. 111–131 in A. Abraham, ed., *The*

Cambridge Handbook of the Imagination. Cambridge, UK: Cambridge University Press, 2020.

Zamani, A., R. Carhart-Harris, and K. Christoff. "Prefrontal Contributions to the Stability and Variability of Thought and Conscious Experience." *Neuropsychopharmacology* 47 (2022): 329–348. PMC8616944.

Chapter 23

Buckner, R. L., and D. C. Carroll. "Self-Projection and the Brain." *Trends in Cognitive Sciences* 11 (2007): 49–57.

Conway, M. A. "Episodic Memories." *Neuropsychologia* 47 (2009): 2305–2313.

Dafni-Merom, A., and S. Arzy. "The Radiation of Autonoetic Consciousness in Cognitive Neuroscience: A Functional Neuroanatomy Perspective." *Neuropsychologia* 143 (2020): 107477.

Fleming, S. M., E. J. van der Putten, and N. D. Daw. "Neural Mediators of Changes of Mind about Perceptual Decisions." *Nature Neuroscience* 21 (2018): 617–624. PMC5878683.

Loftus, E. F. "Resolving Legal Questions with Psychological Data." *American Journal of Psychology* 46 (1991): 1046–1048.

Metcalfe, J., and L. K. Son. "Anoetic, Noetic and Autonoetic Metacognition." In M. Beran et al., eds., *The Foundations of Metacognition*. Oxford, UK: Oxford University Press, 2012.

Miyamoto, K., "Identification and Disruption of a Neural Mechanism for Accumulating Prospective Metacognitive Information Prior to Decision-Making." *Neuron* 109 (2021): 1396–1408. PMC8063717.

Schacter, D. *The Seven Sins of Memory*. Boston: Houghton-Mifflin, 2001.

Schacter, D., and D. Addis. "Memory and Imagination: Perspectives on Constructive Episodic Simulation." Pp. 111–131 in A. Abraham, ed., *The Cambridge Handbook of the Imagination*. Cambridge, UK: Cambridge University Press, 2020.

Suddendorf, T., J. Redshaw, and A. Bulley. *The Invention of Tomorrow*. New York: Basic Books, 2022.

Tulving, E. *Elements of Episodic Memory*. New York: Oxford University Press, 1983.

———. "Episodic Memory and Autonoesis: Uniquely Human?" Pp. 4–56 in H. S. Terrace and J. Metcalfe, eds., *The Missing Link in Cognition*. New York: Oxford University Press, 2005.

———. "The Origin of Autonoesis in Episodic Memory." Pp. 17–34 of H. L. Roediger et al., eds., *The Nature of Remembering: Essays in Honor of Robert G Crowder.* Washington, DC: American Psychological Association, 2001.

Wheeler, M. A., D. T. Stuss, and E. Tulving. "Toward a Theory of Episodic Memory: The Frontal Lobes and Autonoetic Consciousness." *Psychological Bulletin* 121 (1997): 331–354.

Chapter 24

Damasio, A. *Descartes's Error: Emotion, Reason, and the Human Brain.* New York: Gosset / Putnam, 1994.

Fanselow, M. S., and Z. T. Pennington. "The Danger of LeDoux and Pine's Two-System Framework for Fear." *American Journal of Psychiatry* 174 (2017): 1120–1121.

Festinger, L. *A Theory of Cognitive Dissonance.* Evanston, IL: Row Peterson, 1957.

Fleming, S. M. *Know Thyself: The Science of Self-Awareness.* New York: Basic Books, 2021.

Gallagher, S., and D. Zahavi. "Phenomenological Approaches to Self-Consciousness." In E. N. Zalta, ed., *The Stanford Encyclopedia of Philosophy,* Metaphysics Research Lab, Stanford University, 2021. https://plato.stanford.edu/archives/spr2021/entries/self-consciousness-phenomenological.

Girn, M., and K. Christoff. "Expanding the Scientific Study of Self-Experience with Psychedelics." *Journal of Consciousness Studies* 25 (2019): 131–154.

James, W. *Principles of Psychology.* New York: Holt, 1890.

Klein, S. B. "The Feeling of Personal Ownership of One's Mental States: A Conceptual Argument and Empirical Evidence for an Essential, but Underappreciated, Mechanism of Mind." *Psychology of Consciousness: Theory, Research, and Practice* 2 (2015): 355–376.

Koriat, A. "Metacognition and Consciousness." In E. Thompson et al., eds., *The Cambridge Handbook of Consciousness.* Cambridge, UK: Cambridge University Press, 2007.

Lane, T. "Toward an Explanatory Framework for Mental Ownership." *Phenomenology and the Cognitive Sciences* 11 (2012): 251–286.

Lau, H., M. Michel, J. E. LeDoux, and S. M. Fleming. "The Mnemonic Basis of Subjective Experience." *Nature Reviews Psychology* 1 (2022): 479–488. https://doi.org/10.1038/s44159-022-00068-6.

LeCun, Y., Y. Bengio, and G. Hinton. "Deep Learning." *Nature* 521 (2015): 436–444.

LeDoux, J. E. "Coming to Terms with Fear." *Proceedings of the National Academy of Sciences USA* 111 (2014): 2871–2878.

———. "Rethinking the Emotional Brain." *Neuron* 73 (2012): 653–676.

LeDoux, J. E., and H. Lau. "Seeing Consciousness through the Lens of Memory." *Current Biology* 30 (2020): R1018–R1022.

Mangan, B. "The Conscious 'Fringe': Bringing William James Up to Date." Pp. 741–759 in B. J. Baars et al., eds., *Essential Sources in the Scientific Study of Consciousness.* Cambridge, MA: MIT Press, 2003.

Metcalfe, J., and L. K. Son. "Anoetic, Noetic and Autonoetic Metacognition." In M. Beran et al., eds., *The Foundations of Metacognition.* Oxford, UK: Oxford University Press, 2012.

Reber, A. S. "Implicit Learning and Tacit Knowledge." *Journal of Experimental Psychology: General* 118 (1989): 219–235.

Tulving, E. "Episodic Memory and Autonoesis: Uniquely Human?" Pp. 4–56 in H. S. Terrace and J. Metcalfe, eds., *The Missing Link in Cognition.* New York: Oxford University Press, 2005.

———. "The Origin of Autonoesis in Episodic Memory." Pp. 17–34 in H. L. Roediger et al., eds., *The Nature of Remembering: Essays in Honor of Robert G Crowder.* Washington, DC: American Psychological Association, 2001.

Vandekerckhove, M., and J. Panksepp. "A Neurocognitive Theory of Higher Mental Emergence: From Anoetic Affective Experiences to Noetic Knowledge and Autonoetic Awareness." *Neuroscience and Biobehavioral Reviews* 35 (2011): 2017–2025.

Winkielman, P., M. Ziembowicz, and A. Nowak. "The Coherent and Fluent Mind: How Unified Consciousness Is Constructed from Cross-Modal Inputs via Integrated Processing Experiences." *Frontiers in Psychology* 6 (2015): 83. PMC4327174.

Zamani, A., R. Carhart-Harris, and K. Christoff. "Prefrontal Contributions to the Stability and Variability of Thought and Conscious Experience." *Neuropsychopharmacology* 47 (2022): 329–348. PMC8616944.

Chapter 25

Birch, J., A. K. Schnell, and N. S. Clayton. "Dimensions of Animal Consciousness." *Trends in Cognitive Sciences* 24 (2020): 789–801. PMC7116194.

Birch, J., et al. "How Should We Study Animal Consciousness Scientifically?" *Journal of Consciousness Studies* 29 (2022): 8–28.

Browning, H., and J. Birch. "Animal Sentience." *Philosophical Compass* 17 (2022): e12822. PMC9285591.

Chittka, L. *The Mind of a Bee.* Princeton, NJ: Princeton University Press, 2022.

Crump, A., and J. Birch. "Separating Conscious and Unconscious Perception in Animals." *Learning & Behavior* 49 (2021): 347–348.

Dennett, D. C. *Consciousness Explained.* Boston: Little, Brown, 1991.

Gazzaniga, M. S. *The Consciousness Instinct: Unraveling the Mystery of How the Brain Makes the Mind.* New York: Farrar, Straus, and Giroux, 2018.

Kennedy, J. S. *The New Anthropomorphism.* New York: Cambridge University Press, 1992.

Key, B. "Fish Do Not Feel Pain and Its Implications for Understanding Phenomenal Consciousness." *Biology and Philosophy* 30 (2015): 149–165. PMC4356734.

Knoll, E. "Dogs, Darwinism, and English Sensibilities." Pp. 12–21 in R. W. Mitchell et al., eds., *Anthropomorphism, Anecdotes, and Animals.* Albany: State University of New York Press, 1997.

LeDoux, J. E. "As Soon as There Was Life, There Was Danger: The Deep History of Survival Behaviours and the Shallower History of Consciousness." *Philosophical Transactions of the Royal Society London B Biological Sciences* 377 (2022): 20210292. PMC8710881.

———. "What Emotions Might Be Like in Other Animals." *Current Biology* 31 (2021): R824–R829.

Lewis, M. *The Rise of Consciousness and the Development of Emotional Life.* New York: Guilford Press, 2014.

Merker, B. "Consciousness without a Cerebral Cortex: A Challenge for Neuroscience and Medicine." *Behavioral Brain Science* 30 (2007): 63–81, 81–134.

Chapter 26

Baddeley, A. D., R. J. Allen, and G. J. Hitch. "Binding in Visual Working Memory: The Role of the Episodic Buffer." *Neuropsychologia* 49 (2011): 1393–1400.

Barthes, R. "The Death of the Author." In Richard Howard, trans., *The Rustle of Language.* Berkeley: University of California Press, 1986.

Bayne, T. "On the Axiomatic Foundations of the Integrated Information Theory of Consciousness." *Neuroscience of Consciousness* (June 29, 2018): niy007. PMC6030813.

Bruner J. "Life as Narrative." *Social Research* 54 (1987): 11–32.

———. "The 'Remembered' Self." Pp. 41–54 in R. Fivush and U. Neisser, eds., *The Remembering Self: Construction and Accuracy in the Self-Narrative.* Cambridge, UK: Cambridge University Press, 1994.

Campbell, D. T. "'Downward Causation' in Hierarchically Organized Biological Systems." Pp. 176–186 in F. Ayala et al., ed., *Studies in the Philosophy of Biology: Reduction and Related Problems.* London: Macmillan, 1974.

Caruso, G. *Free Will and Consciousness.* Lanham, MD: Lexington Books, 2012.

Dennett, D. C. *Consciousness Explained.* Boston: Little, Brown, 1991.

Dennett, D. C., and G. D. Caruso. *Just Deserts: Debating Free Will.* Medford, MA: Polity Press, 2021.

Einstein, G., and O. Flanagan. "Sexual Identities and Narratives of Self." Pp. 209–231 in G. D. Fireman et al., eds., *Narrative and Consciousness: Literature, Psychology and the Brain.* New York: Oxford University Press, 2003.

Fodor, J. *The Language of Thought.* Cambridge, MA: Harvard University Press, 1975.

Frankland, S. M., and J. D. Greene. "Concepts and Compositionality: In Search of the Brain's Language of Thought." *Annual Review of Psychology* 71 (2020): 273–303.

Gazzaniga, M. S. *Who's in Charge? Free Will and the Science of the Brain.* New York: Harper Collins, 2012.

Gottschall, J. *The Storytelling Animal: How Stories Make Us Human.* New York: Houghton Mifflin Harcourt, 2013.

Lotka, A. J. *Elements of Physical Biology.* Baltimore, MD: Williams and Wilkins, 1925.

Nisbett, R. E., and T. D. Wilson. "Telling More Than We Can Know: Verbal Reports on Mental Processes." *Psychological Review* 84 (1977): 231–259.

Quilty-Dunn, J., N. Porot, and E. Mandelbaum. "The Best Game in Town: The Re-Emergence of the Language of Thought Hypothesis across the Cognitive Sciences." *Behavioral and Brain Sciences* (December 6, 2022): 1–55.

Rescorla, M. "The Language of Thought Hypothesis." In E. N. Zalta, ed., *The Stanford Encyclopedia of Philosophy,* Metaphysics Research Lab, Stanford University, 2019. https://plato.stanford.edu/archives/sum2019/entries/language-thought, accessed April 11, 2022.

Rolls, E. T. "Emotion, Higher-Order Syntactic Thoughts, and Consciousness." Pp. 131–167 in L. Weiskrantz and M. Davies, eds., *Frontiers of Consciousness: Chichele Lectures.* Oxford, UK: Oxford University Press, 2008.

Schneider, S. *The Language of Thought: A New Direction.* Cambridge, MA: MIT Press, 2011.

Searle, J. R. "Minds, Brains, and Programs." *Behavioral and Brain Sciences* 3 (1980): 417–424.

Vygotsky, L. S. *Thinking and Speech: The Collected Works of Lev Vygotsky,* vol. 1. 1934; New York: Plenum, 1987.

Wearing, D. *Forever Today: A Memoir of Love and Amnesia.* London: Doubleday, 2005.

Index

Index